T0182077

Springer Optimization and Its Applications

Volume 140

Aims and Scope
Optimization has been expanding in all directions at an astonishing rate during the last few decades. New algorithmic and theoretical techniques have been developed, the diffusion into other disciplines has proceeded at a rapid pace, and our knowledge of all aspects of the field has grown even more profound. At the same time, one of the most striking trends in optimization is the constantly increasing emphasis on the interdisciplinary nature of the field. Optimization has been a basic tool in all areas of applied mathematics, engineering, medicine, economics and other sciences.

The series *Springer Optimization and Its Applications* publishes undergraduate and graduate textbooks, monographs and state-of-the-art expository works that focus on algorithms for solving optimization problems and also study applications involving such problems. Some of the topics covered include nonlinear optimization (convex and nonconvex), network flow problems, stochastic optimization, optimal control, discrete optimization, multi-objective programming, description of software packages, approximation techniques and heuristic approaches.

More information about this series at http://www.springer.com/series/7393

Ilias S. Kotsireas • Anna Nagurney
Panos M. Pardalos

Editors

Dynamics of Disasters

Algorithmic Approaches and Applications

 Springer

Editors
Ilias S. Kotsireas
Department of Physics & Computer Science
Wilfrid Laurier University
Waterloo, ON, Canada

Panos M. Pardalos
Industrial and Systems Engineering
University of Florida, Center for
Applied Optimization
Gainesville, FL, USA

Anna Nagurney
Department of Operations
and Information Management
Isenberg School of Management
University of Massachusetts
Amherst, MA, USA

ISSN 1931-6828   ISSN 1931-6836   (electronic)
Springer Optimization and Its Applications
ISBN 978-3-030-07356-5   ISBN 978-3-319-97442-2   (eBook)
https://doi.org/10.1007/978-3-319-97442-2

Mathematics Subject Classification: 90, 91, 65, 93

This Springer imprint is published by the registered company Springer Nature Switzerland AG
The registered company address is: Gewerbestrasse 11, 6330 Cham, Switzerland

Preface

This volume is a collection of carefully reviewed papers presented at the 3rd International Conference on Dynamics of Disasters held in Kalamata, Greece, July 5–9, 2017, with additional invited papers. The conference was organized by Ilias S. Kotsireas, Anna Nagurney, and Panos M. Pardalos and convened disaster researchers to discuss their latest scientific work. This volume of 8 papers is organized alphabetically by the first initial of the last name of the first author of each paper with highlights of each paper given below.

The co-editors of this volume are very grateful to the authors of the papers that appear in this volume and also acknowledge the referees for their valuable reports on the papers.

Omkar Achrekar and Chrysafis Vogiatzis in the first paper in this volume, "Evacuation Trees with Contraflow and Divergence Considerations," tackle the important problem of evacuation planning, in the face of an upcoming disaster, in which vehicles utilize the available transportation road network in order to reach safe locations in the form of shelters. They emphasize that, in order to successfully evacuate vehicles located in danger zones, the evacuation process needs to be fast, safe, and seamless. They propose an integer programming model that is based on the concept of an evacuation tree, in which they allow for two policies: contraflow, consisting of lane and street reversals to allow for a higher level of vehicular flow, and divergence, in which evacuees can diverge from the tree under certain conditions. According to the definition of an evacuation tree, vehicles are only allowed to follow one path to safety at each intersection. The authors illustrate the applicability of their framework on two networks, based on the Sioux Falls network and on the Chicago network, accompanied by extensive computational testing and sensitivity analysis, considering both different cost functions and budgets. The solution of their model reveals that it is possible to achieve faster network clearance when utilizing more resources. In addition, after a certain budget limit on the number of contraflows or divergences, the evacuation does not become any faster.

Fuad Aleskerov and Sergey Demin in their paper, "Modelling Possible Oil Spills in the Barents Sea," discuss the rapid increase of the oil and gas industry growth in the Barents Sea during the last few years. They point out that while the Arctic zone is considered to be a relatively clean area, there is a certain number of "hot spots" in the Arctic due to the activities of various companies. They study the problems connected with production of two types of fossil fuel and carried out a simulation model. This model shows the results of oil or gas flowing accident related to drilling complex, taking into account sea currents. By using this model, they can highlight areas in the Barents Sea with the highest potential of the disaster so that preventive measures could be taken. In addition, this model helps to organize elimination of fossil fuel flowing consequences.

Ioanna Falagara Sigala and Fuminori Toyasaki in their paper in this volume, "Prospects and Bottlenecks of Reciprocal Partnerships Between the Private and Humanitarian Sectors in Cash Transfer Programming for Humanitarian Response," discuss the state of the art in practice and in research regarding Cash Transfer Programming (CTP) as an alternative to commodity-based programming, that is, in-kind aid for disaster relief. CTP, in contrast to in-kind aid, transfers purchasing power directly to beneficiaries in the form of currency, which allows them to procure goods and/or services from local markets. In CTP, the private sector, and, especially financial service providers (FSPs), which are entities that provide financial services, including e-transfer services, are essential. FSPs can include e-voucher companies and financial institutions, such as banks and microfinance institutions. Falagara Sigala and Toyasaki, in their study, which is based on primary and secondary qualitative data, present the main characteristics and the mechanisms of CTP to explore how the private sector is involved, identify bottlenecks of reciprocal relationships between financial service providers and humanitarian organizations in CTP, and unveil challenges for the private and humanitarian sectors, which hinder its implementation. This is the first study to explore the partnership of the private sector and the humanitarian sector in CTP, and also describes promising avenues for future research, including new trends in financial donation flows.

Lina Mallozzi and Roberta Messalli in their paper, "Equilibrium Analysis for Common-Pool Resources," present an aggregative normal form game to describe investment decision making for Common-Pool Resources (CPR). A CPR is a natural or human-made resource, from which a group of individuals can benefit, such as, for example, open-seas fisheries. A problem with which a CPR copes is overuse. Indeed, since a CPR is a subtractable resource, that is, its supply is limited, if the quantity that can be restored is overused then there will be a shortage of it, which can result in the destruction of the CPR. The authors consider two directions in order to solve such a problem: the non-cooperative one, characterized by a Nash equilibrium, and the cooperative one, in the form of a fully cooperative equilibrium. They provide existence results for both and, in the case of Environmental Economics, the authors compute and compare the equilibria and then introduce a threshold investment and study the resulting game with aggregative uncertainty. Since an immense range of environmental problems, such as climate change, the loss of biodiversity, ozone depletion, the widespread dispersal of persistent pollutants, and many others, involves the commons, such as forests, energy, industries, water, and so on, this paper has numerous applications.

Anna Nagurney in her paper, "A Multitiered Supply Chain Network Equilibrium Model for Disaster Relief with Capacitated Freight Service Provision," constructs a multitiered game theory model in which the cost-minimizing behavior of disaster relief organizations is captured and that of competing, capacitated freight service providers who are contracted to deliver relief item supplies to multiple points of demand. The governing equilibrium conditions are formulated as a variational inequality problem and existence results provided. The algorithmic approach fully exploits the network structure of the problem and is applied to a timely case study focusing on delivering personal protective equipment to medical professionals battling Ebola in western Africa. The case study reveals that humanitarian organizations benefit from a larger number of competitive freight service providers (although this affects freight service providers negatively in terms of profits). Also, the addition of humanitarian organizations competing for services from the freight service providers results in higher prices since the capacities may be achieved. The paper adds to the literature on game theory and disaster relief, which has only minimally been explored.

Anna Nagurney, Patrizia Daniele, Emilio Alvarez Flores, and Valeria Caruso in their paper, "A Variational Equilibrium Network Framework for Humanitarian Organizations in Disaster Relief: Effective Product Delivery Under Competition for Financial Funds," present a novel Generalized Nash Equilibrium network model, which integrates competition for financial funds among disaster relief organizations and their logistical response post-disasters to provide needed supplies to victims. The humanitarian organizations are subject to common, that is, shared constraints, consisting of lower and upper bounds on the delivered supplies at points of demand with the former guaranteeing that the victims' needs will be met and with the latter constraints reducing material convergence and congestion. The network model is formulated and solved as a variational inequality problem, using a recently introduced concept of a Variational Equilibrium. Lagrange analysis is then utilized to investigate qualitatively the humanitarian organizations' marginal utilities if and when the equilibrium relief item flows are (or are not) at the imposed demand point bounds. The proposed algorithm yields closed form expressions, at each iteration, for the product flows and the Lagrange multipliers, and is applied to a case study, inspired by rare tornadoes that caused devastation in parts of Massachusetts in 2011. The solution to the Nash equilibrium counterparts of the examples making up the case study, in which the common demand bound constraints are removed, is also presented and demonstrates that victims may not receive the required amounts of supplies, without the imposition of the demand bounds. These results provide further support for the need for greater coordination in disaster relief, as made possible by the new model, and show that, by delivering the required amounts of supplies, the humanitarian organizations can also acquire more financial donations, creating a win-win situation.

Ladimer S. Nagurney in his paper, "Advances in Disaster Communications: Broadband Systems for First Responders," begins by overviewing technological changes to public safety disaster communications over the past 90 years from the initial use of wired and simple two-way radios to the advanced broadband systems available to today's first responders. He illustrates how the information and communications needs of first responders have evolved and how coordination and cooperation among various agencies is not the norm. He highlights the challenges and opportunities faced by first responders when not only traditional radio systems are available, but the responders have the ability to use a variety of smart devices connected together through a broadband infrastructure. In this paper, Nagurney also describes the First Responder Network, FirstNet, which is a robust broadband infrastructure in the United States specifically tailored to the needs of first responders. He delineates how a public-private partnership is used to implement FirstNet and reviews the extensions to current broadband technology that will enhance its usefulness to first responders. The author reviews complementary technologies and how they may be used in conjunction with FirstNet. Since first responder use of broadband is not just an American issue, he also discusses current and planned public safety broadband networks across the globe. Nagurney concludes his paper by listing open-ended questions that still need to be solved, including the premier one, is the political/business model sustainable?

Papadaki et al. in the final paper in this volume, "A Humanitarian Logistics Case Study for the Intermediary Phase Accommodation Center for Refugees and Other Humanitarian Disaster Victims," are discussing the recent refugee crisis in Europe as an example of issues arising from the forced mass movement of populations. Some aspects of the theoretical background include historical data regarding the displacement of populations in the European region from the nineteenth century onwards together with the underlying political and economical causes, statistical data highlighting the characteristics and particularities of the current refugee wave and indicating the possible repercussions these could inflict, and finally, the relevant international, European, and national legislation and policies, as well as their potential shortcomings. They then proceed with the proposal of an accommodation center project constructed from modified shipping containers to function as one of the initial stages in adaptation before full social integration of the displaced populations. Some of the key design features include compliance with all current guidelines and regulations regarding space allocations and function, compact overall size suitable even for small plots, and incorporation of several environmental technologies. Regarding the primary function, the accommodation center aims to address significant human needs far beyond food and shelter, such as basic health care, education, administrative assistance, and initiatives for social integration, with the overall goal to maximize the respect for human rights and values while minimizing the impact on society and on the environment. Furthermore, the versatility and specific characteristics of the project make it suitable for any type of humanitarian disaster, thus expanding the original scope significantly and creating a useful tool in all humanitarian relief operations.

Waterloo, ON, Canada   Ilias S. Kotsireas
Amherst, MA, USA   Anna Nagurney[1]
Gainesville, FL, USA   Panos M. Pardalos

[1]Author thanks the Radcliffe Institute for Advanced Study at Harvard University for its support through the Summer Fellowship Program to complete the co-editing of this volume.

Contents

Evacuation Trees with Contraflow and Divergence Considerations

Omkar Achrekar and Chrysafis Vogiatzis

Abstract In this chapter, we investigate the well-studied evacuation planning problem, in which vehicles use the available road transportation network to reach safe areas (shelters) in the face of an upcoming disaster. To successfully evacuate these vehicles located in danger zones, our evacuation process needs to be fast, safe, and seamless. We enable the first two criteria by developing a macroscopic, time-dynamic evacuation model that aims to maximize a reward function, based on the number of people in relatively safer areas of the network at each time point. The third criterion is achieved by constructing an evacuation tree, where vehicles are evacuated using a single path to safety at each intersection. We extend on the definition of the evacuation tree by allowing for divergence and contraflow policies. Divergence enables specific nodes to diverge their flows into two or more streets, while contraflow allows certain streets to reverse their direction, effectively increasing their capacity. We investigate the performance of these policies in the evacuation networks obtained, and present results on two benchmark networks of Sioux Falls and Chicago.

Introduction

Evacuation planning is an important aspect of societal welfare in certain areas worldwide. An active or imminent disaster has the potential to seriously disrupt or destroy the functions of a community, and lead to human life loss, livestock and wildlife relocation or death, material and financial, as well as environmental

O. Achrekar
Department of Industrial and Manufacturing Engineering, North Dakota State University, Fargo, ND, USA
e-mail: omkar.achrekar@ndsu.edu

C. Vogiatzis (✉)
Department of Industrial and Systems Engineering, North Carolina A&T State University, Greensboro, NC, USA
e-mail: cvogiatzis@ncat.edu

© Springer Nature Switzerland AG 2018
I. S. Kotsireas et al. (eds.), *Dynamics of Disasters*, Springer Optimization and Its Applications 140, https://doi.org/10.1007/978-3-319-97442-2_1

damages. These damages are often so big that they exceed the capabilities of modern societies to cope using only their resources [18]. On a different level, that of planning, imminent disasters pose operational challenges on local governments and evacuation managers: the decisions that need to be made at very short notice include whether to evacuate or shelter in place (see, e.g., [8] and the website at [24]), where to locate the necessary personnel and equipments [23], how to best utilize the available resources (see the recent work in [15]), and, of course, how people should evacuate and where they should evacuate to [2]. Our work focuses on this last part.

In this last year, we experienced several disasters that warranted a full-scale evacuation (Hurricane Matthew) or caused crippling damages to the infrastructure (Hurricane Maria). More specifically residents of the United States living along the gulf of Mexico and the Atlantic ocean coast are always asked to have a plan for evacuation and have to carefully monitor weather conditions during the so-called "hurricane season." A well-designed and well-executed plan will not only allow these people to protect themselves, but will also allow communities to bounce back faster. According to the United States Federal Emergency Management Agency (FEMA), the number of disasters that require the residents to evacuate has grown to a staggering 45 to 75 annually [34]. Effective traffic planning and routing is listed at the top as far as the necessary capabilities to ensure mass evacuation [7]. Evacuation planning is critical: in the face of a disaster and without adequate guidance, conditions will only become more chaotic leading to more damages. As a simple example, consider an evacuation plan that suddenly schedules people to use the same street resulting in a sudden overflow of that particular resource and causing congestion.

This chapter is organized as follows. In section Literature Review, we provide relevant literature on the well-studied evacuation problem. We specifically focus on optimization techniques for designing an evacuation plan. Section Mathematical Formulation presents all necessary notation, our assumptions, and the integer programming model we used to devise our evacuation plan. Our model is based on the idea of an *evacuation tree*, in which we allow for two policies, namely *contraflow* (lane and street reversals to enable higher vehicular flow) and *divergence* (where people can diverge from the tree under certain conditions). Then, in section Computational Results, we present our findings on two networks, a small-scale one, based on Sioux Falls, and a larger one, based on the city of Chicago. We conclude our work in section Conclusions.

Literature Review

Evacuation is a very broad term that has spawned very broad, highly interdisciplinary research. Albeit its definition is "moving people from a hazardous area to a safe area," it has numerous aspects ranging from mathematical modeling and solution techniques for large-scale evacuation problems to its behavioral aspects. In this section, we will provide a brief literature review, focusing mainly on its modeling and optimization aspects.

Assumptions and Goals

According to [14], evacuation is the movement of residents from a given area that is considered to be in danger, to other areas that have been designated as safe zones. This process has to be undertaken *quickly*, *reliably*, and *seamlessly*. Evacuations can be done in anticipation of a disaster and as a precautionary measure, but in many cases it is happening during the disaster. Moreover, evacuations can take place in larger (metropolitan areas) and smaller (building) geographical areas. Evacuation modeling can also be divided into macroscopic and microscopic models. In this work, we will focus on anticipatory evacuation that is performed in an urban area using the existing city transportation infrastructure using a macroscopic perspective. Phases of an evacuation process from identification of a threat to arrival of evacuees in a safe zone are described in [32]: our work fits within Phases IV and V of this framework, namely deciding the areas and populations that are affected and routing them to safety.

Relevant Work

Evacuation has attracted significant scientific interest due to its universal nature and its implications for network optimization and graph theory. Evacuation modeling, then, is based on the traffic assignment models (static and dynamic) developed in [3, 5, 6, 25, 29] and a series of network flow problems [9–11, 16, 17]. In the remainder of this subsection, we will discuss contraflow operations and incorporating preferences in the evacuation process.

The concept of using "wasted" capacity from streets leading towards a disaster (instead of leading to safety) is referred to as contraflow [39]. Under such a policy, a lane or street that is underutilized or unused is reversed in direction to increase the capacity of the reverse street and accommodate the increased vehicular flow. Contraflow is no panacea, though, since there are liability issues involved with changing the direction of a street during an emergency situation. For example, during the evacuation of Hurricane Floyd in 1996, the Florida state government did not opt to allow for contraflow. On the other hand, the states of South Carolina and Georgia have now fully incorporated contraflow in their evacuation plans during hurricanes [31]. A review of state practices (including the adoption of contraflow) was published in 2003 [36].

In the literature, contraflow has been shown to render the underlying problem NP-hard [20, 28]. In [19], a contraflow strategy is incorporated in evacuation planning under budget constraints, which appear due to the administrative and operational costs with reversing a street. The authors then solve the resulting mixed integer program and perform a sensitivity analysis, which reaches an interesting result: after a certain "threshold budget," increases in the budget do not yield tangible benefits for the evacuation clearance time. In [37, 38], contraflow is also incorporated within a mixed integer programming framework that is solved using decomposition heuristic approaches. Other work using contraflow includes the tabu-search based heuristic approach in [35], the greedy heuristic in [21], the two heuristics (greedy and combinatorial) presented in [20], and the maximum dynamic contraflow problem setting [26, 27], among others.

An evacuation path needs not only be fast, but also intuitive to the evacuees. One way to enable this is by using the definition of an "evacuation tree" (which we discuss in the next paragraph). Another option would be to only assign evacuees to the nearest safe area (shelter), even if it is not optimal for the whole process. This assignment of evacuees to their closest shelter is adopted in [4, 30].

In an evacuation setting, conflicts of priority in the streets are a serious concern. When multiple vehicles arrive in the same intersection from more than one street, and especially in the case where power is not available and hence the traffic lights are not functional, who leaves the intersection first? One attempt to address this issue was made (over a time-static network) by Cova and Johnson in [4]. In their work, they minimize the conflicts that take place in any intersection. Another work that has been instrumental in minimizing conflicts in intersections and ensuring the seamlessness of the evacuation process is based on the concept of an "evacuation tree," introduced in [1]. A small example of an evacuation tree versus an earliest time approach to solving an evacuation problem is given in Figure 1.

Seamless evacuation is achieved using evacuation trees by disallowing people in the same location to evacuate using different paths. Hence, at each node of our transportation network, vehicles have exactly one path to safety. While restrictive, this enables a more organized way to evacuate people from an urban area. In our work, we enrich the evacuation tree approach by incorporating two improving policies (contraflow and divergence) with the goal of evacuating the vehicles from the danger zones of the network faster. Contraflow effectively increases the capacity of the streets through street reversals. Divergence allows for specific nodes to divert their flows into more than one path to safety (and hence spread out the flow on different outgoing streets) under special conditions.

The evacuation tree enforces a restriction on the paths used by the evacuees. It requires all the people from the same source to follow a common path. The idea of tree generation can be compared with heuristic developed by Lu et al. [22] which assigns the evacuees to clusters and are then routed through the network by shortest path search. The difference between these two approaches is that in the clustering technique evacuees at the intermediate node can be routed through the different paths, but with the evacuation tree model we require evacuees entering an intersection to leave that node using only one exit.

It is intuitive to use multiple arcs leaving a node to help evacuees reach safety faster. Using more than one street to leave an intersection can be useful in the case of an emergency and especially at a node which is highly centralized: this means that this node is used by multiple routes to safety. Hence, in our proposed evacuation model, we are allowing some of the tree nodes of the network to diverge further depending upon budget availability.

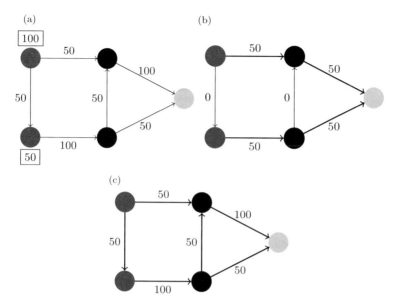

Fig. 1 The nodes in red are in the danger zone and are expected to evacuate in the green area. The capacities in the network (shown in **a**) can accommodate all demand in one time period (send all 150 vehicles to the shelter), but to achieve that, all streets are used and a cycle is formed: vehicle located in the top left red node are routed through the bottom left node and its neighbor, before they get back to the route other evacuees were following (**c**). This is avoided in the evacuation tree (see **c**), but to the expense of needing one more time period to safely evacuate everyone. (**a**) The original network capacities. (**b**) The evacuation tree and its flows per time period. (**c**) Fastest evacuation configuration and its flows per time period

Mathematical Formulation

Notation

Let $G(V, E)$ be the available transportation network, with V representing the set of nodes (intersections) and E the set of arcs (streets) connecting any two nodes. In this work, we are considering a time-expanded network, and hence we deal with an instance of the graph at any time $t = 0, \ldots, T$ where T is the total amount of time available for the evacuation process.

We also assume that our evacuation area is divided into 3 zones. Those zones are denoted by $S_1, S_2, S_3 \subset V$ such that $S_1 \cup S_2 \cup S_3 = V$ and $S_1 \cap S_2 \cap S_3 = \emptyset$. The intuition behind the zones is clear: S_1 (zone 1) is the one that is geographically closest to the disaster origin points and is in immediate danger. S_3 (zone 3), on the other hand, consists of all nodes that have been designated to be safe. Finally, S_2 (zone 2) is an intermediary zone, which connects vehicles located in the danger zone to the safe zone.

Moreover, we assume that each node has an initial number of vehicles (initial population) which are expected to evacuate. A good evacuation policy, then, will try to make sure that all vehicles in the transportation network at time $t = 0$ reach any safe area $i \in S_3$ before the final time step T. Another assumption we make is that nodes have no capacity consideration, whereas arcs have a limited capacity. This capacity (u_{ij}) is also incorporated in the time-expanded network, implying that vehicles traversing the street count towards that upper limit, no matter when they started the traversal. We also assume that every street has a known (and deterministic) travel time ω_{ij}. Last, for every node $i \in S_1 \cup S_2$ (i.e., all the transportation network nodes excluding the safe zone), we assume that there is a *danger factor* r_i^t, which is dependent on the time step t, such that $r_i^{t_1} \geq r_i^{t_2}, \forall t_1 \geq t_2$. For nodes $i \in S_3$, r_i^t instead represents a *safety factor* and is used as a reward in our objective function for reaching a safe area. The danger (resp., safety) factors of all nodes in $S_1 \cup S_2$ (resp., S_3) are assumed to be inputs to our problem, and are used as parameters in our model.

Our work is based on the concept of an evacuation tree, as described in section Relevant Work. Herein though, we allow for two modifications. The first one is based on the notion of contraflow (also introduced in section Relevant Work). We assume that each street can be reversed with a cost of $\bar{c}_{ij}$: of course, when a street is reversed it can no longer be used, as its capacity is added to the one of the opposite direction. The second modification allows for divergences from the tree. One of the main ideas of an evacuation tree (that also establishes its seamlessness and intuitive nature) is that any vehicle at any intersection has exactly one path to safety. Here, we enable certain intersections to diverge flow in more than one path to safety when:

1. A resource (police officer, special equipment) is budgeted to allow this;
2. The number of diverging paths is at most equal to the number of paths entering the intersection.

We provide a small pictorial example of these conditions in Figure 2.

Given a network $G(V, E)$, we define the forward (resp., reverse) star of a node i as the nodes that are reachable from i (resp., the nodes that can reach i). Formally, we will write that $FS(i) = \{j \in V : (i, j) \in E\}$ and $RS(i) = \{j \in V : (j, i) \in E\}$. Last, given a set of nodes S, we say that $G[S]$ with a node set $V_S = S$ and an edge set $E_S = \{(i, j) \in E : i \in S, j \in S\}$ is the induced subgraph of S.

Mathematical Model

Using the definitions from the previous subsection, we can now define the decision variables of our model. The decisions being are (i) the streets to be used, (ii) the streets to be reversed, (iii) the intersections to be diverged, as well as (iv) the flows and demands on streets and intersections, respectively. The full model is shown in (1).

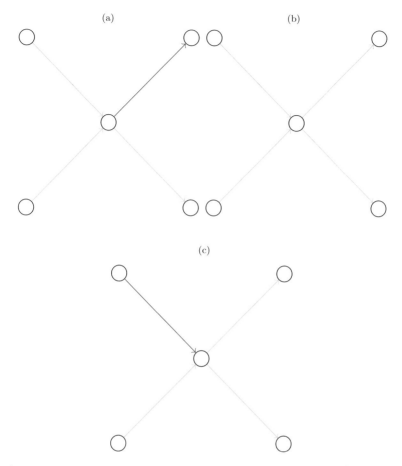

Fig. 2 An example of the divergence conditions. The arcs in green are part of the evacuation plan, while the ones in red are not and do not have vehicle flow. (**a**) Allowed. This is an evacuation tree. (**b**) Allowed. This is not an evacuation tree, but the middle node that diverges has two streets entering and two streets exiting. (**c**) Not allowed. This is not an evacuation tree, and the middle node that diverges has only one street entering but two streets exiting

$$f_{ij}^t = vehicular\ flow\ on\ arc(i,j) \in E\ at\ time\ t = 1, \ldots, T.$$

$$d_i^t = remaining\ demand\ at\ node\ i \in V\ at\ time\ t = 1, \ldots, T.$$

$$x_{ij} = \begin{cases} 1, & \text{if arc } (i,j) \text{ is used in the evacuation plan,} \\ 0, & \text{otherwise.} \end{cases}$$

$$y_{ij} = \begin{cases} 1, & \text{if arc } (i,j) \text{ is reversed in the evacuation plan,} \\ 0, & \text{otherwise.} \end{cases}$$

$$m_i = number\ of\ divergences\ a\ node\ i \in V\ is\ allowed.$$

$$\min \sum_{i \in V} \sum_{t=1}^{T} r_i^t d_i^t \tag{1a}$$

$$s.t. \quad \sum_{\tau = max\{0, t-\omega_{ij}+1\}}^{min\{t, T-\omega_{ij}\}} f_{ij}^\tau \le u_{ij} x_{ij} + u_{ji} y_{ji}, \qquad \forall (i,j) \in E, \forall t = 1, 2, \ldots, T-1,$$
$$\tag{1b}$$

$$\sum_{j \in FS(i)} f_{ij}^t - \sum_{j \in RS(i)} f_{ji}^{t-\omega_{ji}} + d_i^t - d_i^{t-1} = 0, \qquad \forall i \in V, \forall t = 1, 2, \ldots, T-1,$$
$$\tag{1c}$$

$$\sum_{j \in FS(i)} x_{ij} \le \sum_{j \in RS(i)} x_{ji}, \qquad \forall i \in V,$$
$$\tag{1d}$$

$$\sum_{j \in FS(i)} x_{ij} = 1 + m_i, \qquad \forall i \in V,$$
$$\tag{1e}$$

$$\sum_{i \in V} \hat{C}_i m_i \le \hat{B}, \tag{1f}$$

$$\sum_{(i,j) \in E} \bar{C}_{ij} y_{ij} \le \bar{B}, \tag{1g}$$

$$y_{ji} \le x_{ij}, \qquad \forall (i,j) \in E, \tag{1h}$$

$$\sum_{i \in S, j \in S} x_{ij} \le |S| - 1, \qquad \forall S \subset V, \tag{1i}$$

$$f_{ij}^t \ge 0, \qquad \forall (i,j) \in E, \forall t = 1, \ldots, T, \tag{1j}$$

$$d_i^t \ge 0, \qquad \forall i \in V, \forall t = 1, \ldots, T, \tag{1k}$$

$$x_{ij} \in \{0, 1\}, \qquad \forall (i,j) \in E, \tag{1l}$$

$$y_{ij} \in \{0, 1\}, \qquad \forall (i,j) \in E, \tag{1m}$$

$$m_i \in \mathbb{Z}^*, \qquad \forall i \in V. \tag{1n}$$

We are using a dynamic network flow model to represent our evacuation problem. The objective function is presented in (1a) and minimizes the total cost of evacuation. The cost is a penalty incurred by an evacuee by spending time in a danger zone: it is a reward in the case of evacuees that have already reached safety. This penalty changes over time: if an evacuee stays in the danger zone for longer periods, they get penalized more. Similarly, an evacuee reaching safety earlier is rewarded more. Constraint family (1b) ensures that all streets have an upper bound on the amount of flow (number of evacuating vehicles) allowed. In some works, this capacity is changing over time and/or depending on the flow at each time point: in our work, we assume (for simplicity) that it is constant. An arc (i, j) is only allowed to accommodate flow if $x_{ij} = 1$. In the same constraint, if the reverse arc (j, i) is selected for contraflow (i.e., $y_{ji} = 1$), its capacity gets augmented. The flow balance constraints for a dynamic problem are presented in constraints (1c). This constraint family is the time expanded version of the flow balance constraints for time static network flow problems. Further, we make the assumption that all flows are uninterrupted, implying that there are no delays experienced while in transit.

Constraints (1d) and (1e) are to restrict the number of outgoing arcs for every intersection/node. Constraints (1d) limit the number of outgoing arcs to be less than or equal to the number of incoming arcs in that particular node. Constraints (1e) allow outgoing arcs from node i to be at least equal to 1 and, should the model allows that node to diverge, then an integer value for m_i is selected to signal the extra number of arcs that are allowed to exit that node. This number of divergences depends upon the budget availability, which is enforced with constraint (1f). Similarly, constraint (1g) restricts the arcs to be selected for contraflow within the budget limits. Constraints (1h) allow only arcs that have a reverse street selected to carry flow be eligible for contraflow. Constraints (1i) are cycle-breaking constraints to avoid sending vehicles in circles while waiting to be evacuated. We note here that in the work of [1], the cycle-breaking constraints are omitted as it can be shown that no cycles will appear in an optimal solution, which unfortunately does not hold in our work. Finally, (1j)–(1n) are variable restrictions, in accordance with their definition earlier in the subsection.

Computational Results

We obtain all results of our mathematical program under three different cost functions using Gurobi 7.5 [12]. All codes were written in Python, with visualizations being done using NetworkX [13]. All numerical experiments were performed on an Intel Core i7-5500U at 2.39 GHz.

Two benchmark instances (available at https://github.com/bstabler/Transportation-Networks) [33] were analyzed: one is based on the city of Sioux Falls, while the other is a sketch network of Chicago. The first network is visualized in Figure 3, while (due to its size) the second is not. Three sink (safe zone S_3) costs are

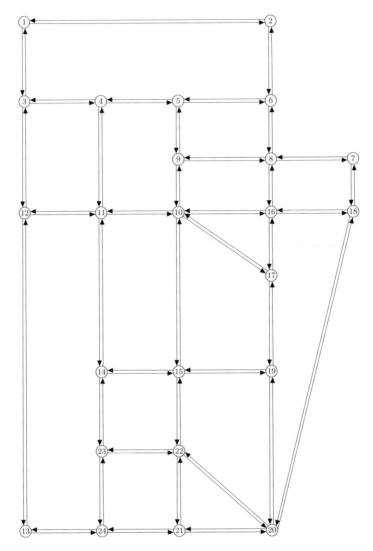

Fig. 3 The transportation network for Sioux Falls, downloaded from https://github.com/bstabler/TransportationNetworks, visualized

considered: $\{-1, -t, t - T\}$. The insight behind each of the costs is as follows. The first cost affects all evacuees the same no matter their time of arrival. The second is dynamic and discounts later arrivals. The last cost is also dynamic and behaves similarly to the previous one, however it more drastically rewards early arrivals to safety. A small example on how these costs behave is provided in Table 1.

Table 1 Consider two vehicles that evacuate at time $t = 2$ and time $t = 4$, when the total evacuation horizon is $T = 6$

Cost	$t = 1$	$t = 2$	$t = 3$	$t = 4$	$t = 5$	$t = 6$
Cost: -1	0	-1	-2	-4	-6	-8
Cost: $-t$	0	-2	-5	-13	-23	-35
Cost: $t - T$	0	-4	-7	-11	-13	-13

From the table, it can be seen that the first cost is essentially a counter of how many periods vehicles have spent to safety (in the example, this is 8 periods – 5 for the first vehicle and 3 for the second). The second cost is dynamic and rewards vehicles that are in safety as much as the discrete time they arrived in (hence the first vehicle gets a reward of 2, 3, 4, 5, and 6 for a total 20, and the second vehicle rewards of 4, 5, and 6 for a total of 15). Finally, the last cost still rewards vehicles that arrive to safety, but discounts late arrivals (the first vehicle now receives rewards of 4, 3, 2, 1, and 0, for a total of 10, and the second vehicle rewards of 2, 1, and 0, for a total of 13).

For simplicity, in some parts of the computational results, we will use the notation abc to represent an availability for a divergences, a budget of b contraflows, and the c-th cost configuration with $c = 1, 2, 3$ for $\{-1, -t, t - T\}$, respectively. In accordance with our assumptions (see section Notation), we are not allowing nodes which belong to S_1 to bifurcate to avoid chaos; lanes and streets with one endpoint to the danger zone though are allowed to be reversed and serve for contraflow.

The next subsections present all evacuation plans obtained for the Sioux Falls network under varying budgets for divergence and contraflow both in tabular and pictorial format. For Chicago, the results are presented only in table format, for ease of presentation purposes.

Sioux Falls

We begin by presenting the three evacuation trees obtained without any contraflows or divergences allowed. This is shown in Figure 4 for all three cost configurations (see Figure 4a–c). We note that the two last costs ($-t$ and $t - T$) produce the same output.

We now proceed to present the results when allowing for 5 and 10 divergences in all three different cost configurations in Figures 5, 6, and 7.

The evacuation networks designed when using all three costs are very similar. The nodes considered for allowing divergences (under each budget allowed) are exactly the same, as shown in Table 2. When comparing our findings to the simple evacuation trees of Figure 4, the benefits of enabling divergences is clear: in Table 3, we see how the total clearance time decreases from 98 periods to as little as 67 periods when allowing for 10 divergences in the evacuation plan. All three costs showcase similar evacuation rates also (Table 4).

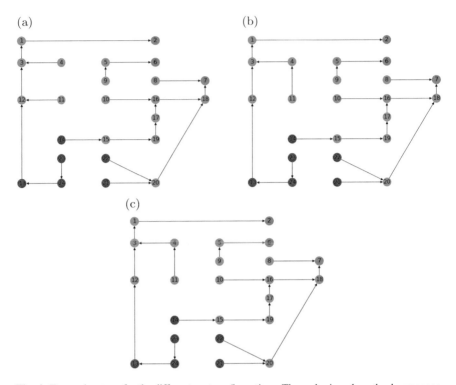

Fig. 4 Evacuation trees for the different cost configurations. The nodes in red are the danger zone, and the nodes in green the safe areas. (**a**) Evacuation tree for cost -1. (**b**) Evacuation tree for cost $-t$. (**c**) Evacuation tree for cost $t - T$

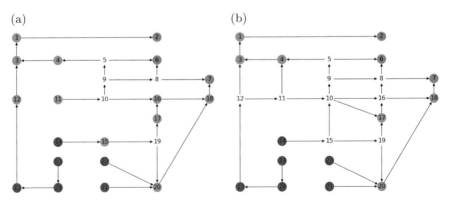

Fig. 5 Evacuation trees for cost configuration -1 under different divergence budgets. (**a**) 5 divergences allowed. (**b**) 10 divergences allowed

As shown in Figure 8, the difference between the three costs in all divergence budgets are indistinguishable. The figure also shows how people reach safety; we note that all the cost configurations result in evacuations of roughly the same rate.

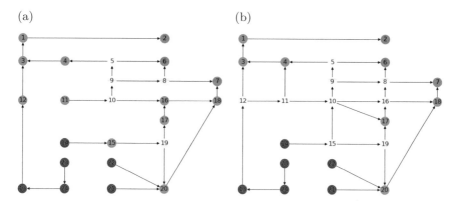

Fig. 6 Evacuation trees for cost configuration $-t$ under different divergence budgets. (**a**) 5 divergences allowed. (**b**) 10 divergences allowed

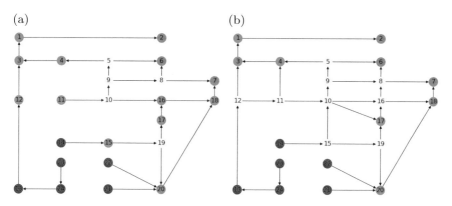

Fig. 7 Evacuation trees for cost configuration $t - T$ under different divergence budgets. (**a**) 5 divergences allowed. (**b**) 10 divergences allowed

Table 2 Nodes selected to be diverged in all evacuation trees under different divergence budgets

Nodes	Divergences allowed	
Selected	5	10
5	x	x
8	x	x
9	x	x
10	x	xx
11		x
12		x
15		x
16		x
19	x	x

Table 3 Total network clearance time when considering divergences only

	Sink costs		
Combination	-1	$-t$	$t-T$
0_0	98	98	98
5_0	72	67	67
10_0	67	67	67

Table 4 Danger zone clearance time when considering divergences only

	Sink costs		
Combination	-1	$-t$	$t-T$
0_0	56	56	56
5_0	56	56	56
10_0	56	56	56

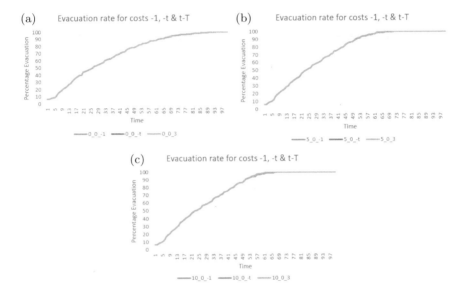

Fig. 8 Network evacuation rates for all plans under different divergence budgets. (**a**) 0 divergence allowed. (**b**) 5 divergences allowed. (c) 10 divergences allowed

However, it is interesting to note that while an improved clearance time is expected, this is not necessarily reflected in the clearance time for the danger zone: with or without a divergence budget, all three costs evacuate the population present at the danger zone in the beginning of the evacuation process at the same time (after 56 periods). We continue with similar results when considering only a budget for contraflows in Figures 9, 10, and 11 (Tables 5, 6, 7, 8, 9).

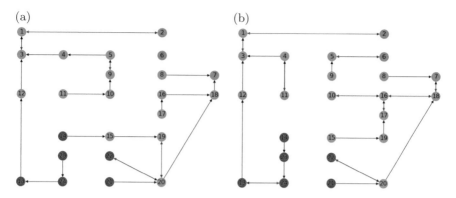

Fig. 9 Evacuation trees for cost configuration −1 under different contraflow budgets. (**a**) 5 contraflows allowed. (**b**) 10 contraflows allowed

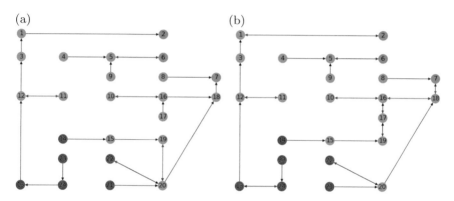

Fig. 10 Evacuation trees for cost configuration −t under different contraflow budgets. (**a**) 5 contraflows allowed. (**b**) 10 contraflows allowed

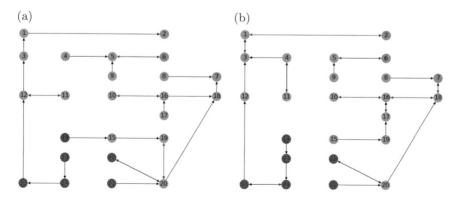

Fig. 11 Evacuation trees for cost configuration $t − T$ under different contraflow budgets. (**a**) 5 contraflows allowed. (**b**) 10 contraflows allowed

Table 5 Arcs selected to be reversed in the evacuation trees for cost configuration -1

Arcs	Contraflows allowed	
Selected	5	10
(1,3)	x	x
(2,1)	x	x
(4,11)		x
(5,9)	x	
(6,5)		x
(7,18)		x
(13,24)		x
(16,10)		x
(16,17)		x
(18,16)		x
(20,19)	x	
(20,22)	x	x

Table 6 Arcs selected to be reversed in the evacuation trees for cost configuration $-t$

Arcs	Contraflows allowed	
Selected	5	10
(2,1)		x
(6,5)	x	x
(7,18)		x
(12,11)	x	x
(13,24)		x
(16,17)		x
(16,10)	x	x
(17,19)		x
(18,16)		x
(20,19)	x	
(20,22)	x	x

Table 7 Arcs selected to be reversed in the evacuation trees for cost configuration $t - T$

Arcs	Contraflows allowed	
Selected	5	10
(1,3)		x
(2,1)		x
(4,11)		x
(6,5)	x	x
(7,18)		x
(12,11)	x	
(13,24)		x
(16,10)	x	x
(16,17)		x
(18,16)		x
(20,19)	x	
(20,22)	x	x

Table 8 Total network clearance time when considering contraflow only

| | Sink costs | | |
Combination	-1	$-t$	$t - T$
0_0	98	98	98
0_5	72	62	62
0_10	54	54	54

Table 9 Danger zone clearance time when considering contraflow only

| | Sink costs | | |
Combination	-1	$-t$	$t - T$
0_0	56	56	56
0_5	39	39	39
0_10	33	31	33

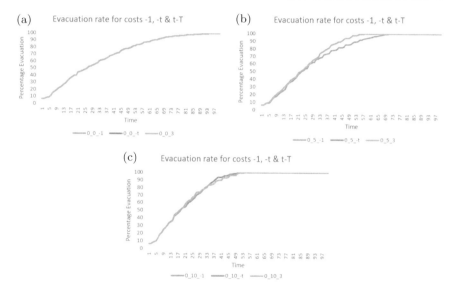

Fig. 12 Network evacuation rates for all plans under different contraflow budgets. (**a**) 0 contraflows allowed. (**b**) 5 contraflows allowed. (**c**) 10 contraflows allowed

Contrary to the divergence case, it is evident from Figures 9, 10, and 11 that the three studied costs produce different evacuation networks: they also consider different arcs to be reversed during the evacuation process (Figure 12).

We are now ready to consider both divergence and contraflow at the same time in Figures 13, 14, and 15. We will sometimes refer to this scheme of allowing both policies at the same time as the *coupled* scheme.

Figures 13, 14, and 15 show the evacuation networks generated by the model when allowing both contraflow and divergence and using costs -1, $-t$, and $t - T$ respectively. Tables 10, 12, and 14 show the lanes used for the contraflows by the model for all three types of costs. According to Table 10 lanes (2,1), (1,3), and lane (5,9) are very important for the cost -1 as these lanes keep appearing

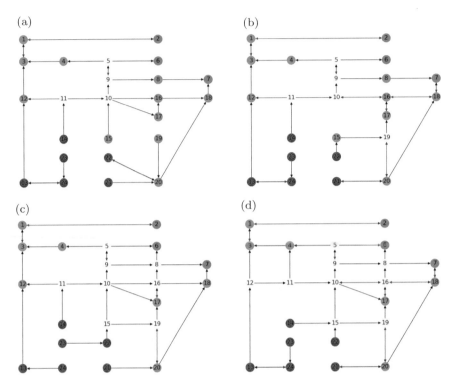

Fig. 13 Evacuation trees for cost configuration -1 under different contraflow and divergence budgets. (**a**) 5 divergences and 5 contraflows allowed. (**b**) 5 divergences and 10 contraflows allowed. (**c**) 10 divergences and 5 contraflows allowed. (**d**) 10 divergences and 10 contraflows allowed

frequently in the combinations. Similarly, lanes (1,3), and (2,1) are important under cost configurations $-t$ and $t - T$. There are some lanes which are only used when contraflows are allowed without any divergences (e.g., arc (4,11) under the -1 and $t - T$ cost configurations). As the budget increases, costs -1 and $-t$ produce exactly the same evacuation networks for 5_10, 10_5 and 10_10 combinations. For divergence, node 10 seems to play a significant role, as it is allowed to diverge even more when the budget allows (in all configurations allowing 10 divergences) (Figures 16, 17, 18, 19, 20; Tables 11, 12, 13, 14).

Finally, we consider the effect that the different cost configurations play on the evacuation rates and clearance times. We observe an interesting outcome for the different cost configurations: when using a sink cost of -1 or t, we see an inconsistency, where for lower budgets, the clearance time is smaller. This happens because vehicles near the danger zone are evacuated faster to receive the "reward" of evacuating them faster and having them stay at the safe area. That said, cost $t - T$ leads to consistent results, as an increase in the budget of contraflows or

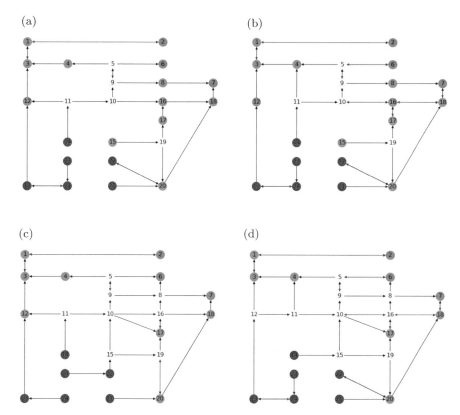

Fig. 14 Evacuation trees for cost configuration $-t$ under different contraflow and divergence budgets. (**a**) 5 divergences and 5 contraflows allowed. (**b**) 5 divergences and 10 contraflows allowed. (**c**) 10 divergences and 5 contraflows allowed. (**d**) 10 divergences and 10 contraflows allowed

divergences allowed leads to an improvement in the evacuation plan, and hence it is a good choice for designing the evacuation tree (Tables 15, 16, 17).

Chicago

The "Chicago Sketch" benchmark network that was used has 933 nodes and 2950 arcs. We again use the same experiment setup as previously with the three cost configurations $(-1, -t, t - T)$. The total horizon for the evacuation process now is $T = 150$.

Our results are presented in Tables 18, 19, 20 (which show the arcs selected for contraflow under the different costs), Tables 21, 22, and 23 (which present the nodes

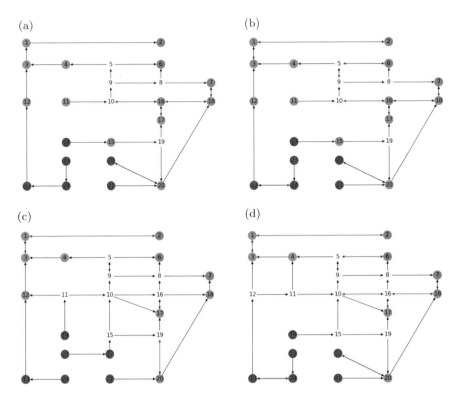

Fig. 15 Evacuation trees for cost configuration $t - T$ under different contraflow and divergence budgets. (**a**) 5 divergences and 5 contraflows allowed. (**b**) 5 divergences and 10 contraflows allowed. (**c**) 10 divergences and 5 contraflows allowed. (**d**) 10 divergences and 10 contraflows allowed

Table 10 Arcs selected to be reversed in the evacuation trees of the coupled scheme for cost configuration -1

Arcs	Contraflows and divergences allowed					
Selected	0_5	0_5	5_5	5_10	10_5	10_10
(1,3)	x	x	x	x	x	x
(2,1)	x	x	x	x	x	x
(4,11)		x				
(5,9)	x		x	x	x	x
(6,5)		x				
(7,18)		x		x		x
(13,24)		x	x	x		x
(16,10)		x		x		x
(16,17)		x		x	x	x
(18,16)		x		x		x
(20,19)	x			x	x	x
(20,21)					x	x
(20,22)	x	x	x			

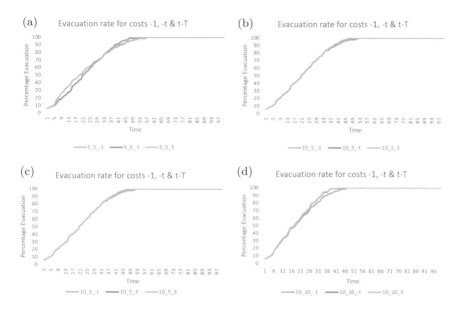

Fig. 16 Network evacuation rates for all plans under different divergence and contraflow budgets. (**a**) 5 divergences and 5 contraflows allowed. (**b**) 5 divergences and 10 contraflows allowed. (**c**) 10 divergences and 5 contraflows allowed. (**d**) 10 divergences and 10 contraflows allowed

used for divergence), and Tables 24, 25, 26, 27, 28, and 29 (for when allowing both contraflow and divergences).

The Chicago benchmark instance is a large-scale network, so we keep our analysis to tabulated results, instead of providing any visual insights on the network, like we did for the first benchmark network.

First, consider Tables 18, 19, and 20, which present our results when allowing only contraflow. It is evident that there is a series of arcs which are used in all three cost configurations, e.g., arcs (424,425), (426,441), (438,535), and (777,778). We can then say that these arcs must be significant when we have a budget allowing only for contraflows as they appear in all the combinations. On a similar note, there exist some arcs which are only important in specific cost configurations, e.g., arc (434,435) is important for the first cost configuration (-1), only appears once in the last configuration $(t - T)$, and is not considered at all for the second cost configuration.

Now, we consider the case where we only have a budget for divergences. According to Tables 21, 22, and 23 we certainly can see some patterns followed by these three costs. Nodes 398, 428, 436, 535, 622, and 902 are used by all three types of the costs for all the budgets available. Once more, there are certain nodes which are only employed for divergence under specific costs. For example, nodes 392, 393 and 413 are used solely by cost $t - T$ and nodes 631, 643, 686, and 696 by cost $-t$, whereas finally node 903 is only used only cost -1. Enabling divergences in the

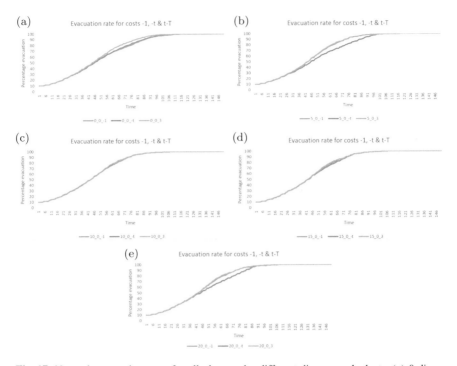

Fig. 17 Network evacuation rates for all plans under different divergence budgets. (**a**) 0 divergences and 0 contraflows allowed. (**b**) 5 divergences and 0 contraflows allowed. (**c**) 10 divergences and 0 contraflows allowed. (**d**) 15 divergences and 0 contraflows allowed. (**e**) 20 divergences and 0 contraflows allowed

evacuation trees provides us with lower total network clearance times as presented in Table 30.

So far we have seen the effects of adding budgets for contraflows and divergences in the evacuation tree. Finally, we consider the case of having both a budget for contraflows and divergences. Here we will go over some other important nodes and arcs used by all the cost configurations. First, we consider Tables 24, 25, 26. Nodes 400, 428, 436 seem to be important for all the costs, since they appear in all combinations. Take a look at node 584: it is not important at the lower budget levels for costs $t - T$ and -1, but the opposite is true for cost $-t$. Node 648 is not very important for cost $-t$, as it appears only twice in all the coupled combinations. For cost -1, it is somewhat important; finally, for cost $t - T$ it is not used at all when the budget is lower, but as the budget increases it appears frequently. Contraflows also share some common arcs for all three cost configurations. Arcs (424,425), (426,441), (431,432), and (777,778) are important arcs for all the three costs and all the coupled combinations. Arc (769,771) and (777,779) are reversed only after the budget for contraflows increases (Table 31).

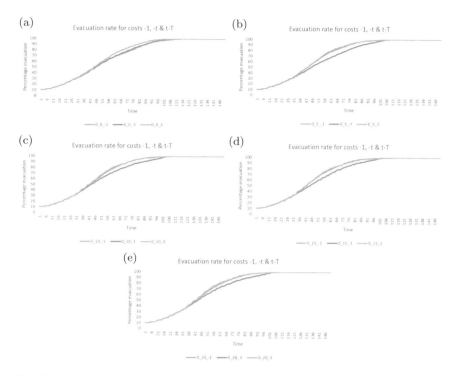

Fig. 18 Network evacuation rates for all plans under different contraflow budgets. (**a**) 0 divergences and 0 contraflows allowed. (**b**) 0 divergences and 5 contraflows allowed. (**c**) 0 divergences and 10 contraflows allowed. (**d**) 0 divergences and 15 contraflows allowed. (**e**) 0 divergences and 20 contraflows allowed

Conclusions

Evacuation planning is a difficult, yet important problem. Especially in larger-scale networks, where the number of people evacuating is very high, a careful balance between fast evacuation to safety and plan seamlessness needs to be achieved. In this work, we proposed a mathematical formulation that is based on the definition of an evacuation tree, where vehicles are only allowed to follow one path to safety at each intersection. This leads to better evacuation organization and decreased coordination costs. At the same time, though, it leaves certain resources (streets, intersections, officers) underutilized. Hence, in our work we add two budgets for contraflow (street reversals) and divergence (allowing for more than one path to safety at certain intersections). We show how the evacuation trees obtained in two benchmark networks, based on the cities of Sioux Falls and Chicago, are changing with increased budgets for this operations.

Using this model, we were able to show that, intuitively, it is possible to achieve faster network clearance when using more resources. It has also been demonstrated that after a certain budget limit (on the number of contraflows or divergences), the

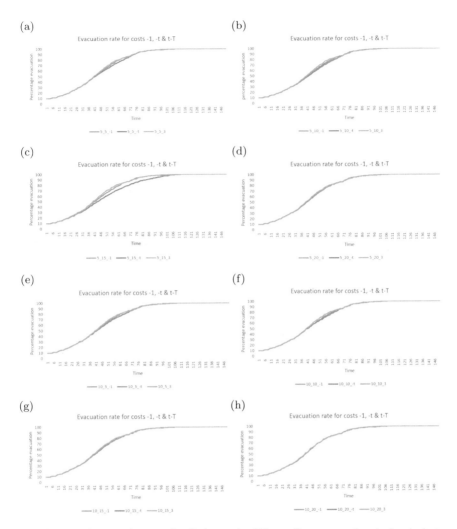

Fig. 19 Network evacuation rates for all plans under different divergence and contraflow budgets. (**a**) 5 divergences and 5 contraflows allowed. (**b**) 5 divergences and 10 contraflows allowed. (**c**) 5 divergences and 15 contraflows allowed. (**d**) 5 divergences and 20 contraflows allowed. (**e**) 10 divergences and 5 contraflows allowed. (**f**) 10 divergences and 10 contraflows allowed. (**g**) 10 divergences and 15 contraflows allowed. (**h**) 10 divergences and 20 contraflows allowed

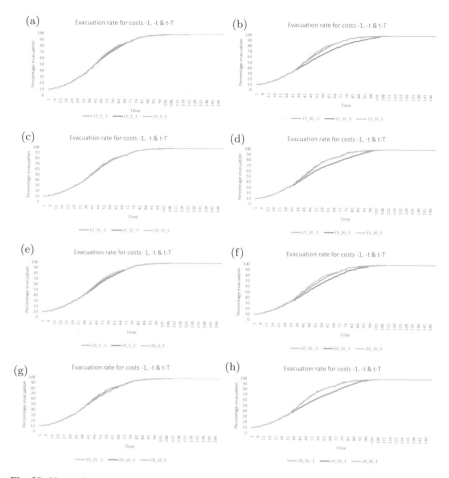

Fig. 20 Network evacuation rates for all plans under different divergence and contraflow budgets. (a) 15 divergences and 5 contraflows allowed. (b) 15 divergences and 10 contraflows allowed. (c) 15 divergences and 15 contraflows allowed. (d) 15 divergences and 20 contraflows allowed. (e) 20 divergences and 5 contraflows allowed. (f) 20 divergences and 10 contraflows allowed. (g) 20 divergences and 15 contraflows allowed. (h) 20 divergences and 20 contraflows allowed

evacuation does not become any faster. Finally, we showed that using a dynamically updated cost of $t - T$ helps to evacuate people in a more consistent manner, as compared to the other two costs we put to the test, equal to -1 (static) and $-t$ (dynamic).

For future work, we can further refine the selection of nodes to diverge at and streets to reverse using graph theoretic notions of centrality. For example, we can allow nodes to diverge if they possess higher betweenness centrality over the paths that leads to safety. Moreover, we can investigate lane reversals (instead of street reversals): that way, certain streets can still have some leftover capacity for emergency vehicles that need access to the danger zone areas, if necessary.

Table 11 Nodes selected to be diverged in the evacuation trees of the coupled scheme for cost configuration -1

Nodes	Contraflows and divergences allowed					
Selected	5_0	10_0	5_5	5_10	10_5	10_10
5	x	x	x	x	x	x
8	x	x			x	x
9	x	x	x	x	x	x
10	x	xx	xx	x	xx	xx
11		x	x	x	x	x
12		x				x
15		x			x	x
16		x			x	x
19	x	x		x	x	x

Table 12 Arcs selected to be reversed in the evacuation trees of the coupled scheme for cost configuration $-t$

Arcs	Contraflows and divergences allowed					
Selected	0_5	0_10	5_5	5_10	10_5	10_10
(1,3)			x	x	x	x
(2,1)		x	x	x	x	x
(5,9)			x	x	x	x
(6,5)	x	x		x		x
(7,18)		x		x		x
(12,11)	x	x				
(13,24)		x	x	x		x
(18,16)		x		x		x
(16,10)	x	x		x		x
(16,17)		x		x	x	x
(17,19)		x				
(20,19)	x				x	
(20,22)	x	x	x	x		x

Table 13 Nodes selected to be diverged in the evacuation trees of the coupled scheme for cost configuration $-t$

Nodes	Contraflows and divergences allowed					
Selected	5_0	10_0	5_5	5_10	10_5	10_10
5	x	x	x	x	x	x
8	x	x			x	x
9	x	x	x	x	x	x
10	x	xx	x	x	xx	xx
11		x	x	x	x	x
12		x				x
15		x			x	x
16		x			x	x
19	x	x	x	x	x	x

Table 14 Arcs selected to be reversed in the evacuation trees of the coupled scheme for cost configuration $t - T$

Arcs Selected	Contraflows and divergences allowed					
	0_5	0_10	5_5	5_10	10_5	10_10
(1,3)		x		x	x	x
(2,1)		x		x	x	x
(4,11)		x				
(5,9)				x	x	x
(6,5)	x	x		x		x
(7,18)		x	x	x		x
(12,11)	x					
(13,24)		x		x		x
(16,10)	x	x	x	x		x
(16,17)		x	x	x	x	x
(18,16)		x	x	x		x
(20,19)	x				x	
(20,22)	x	x	x	x		x

Table 15 Nodes selected to be diverged in the evacuation trees of the coupled scheme for cost configuration $t - T$

Nodes Selected	Contraflows and divergences allowed					
	5_0	10_0	5_5	5_10	10_5	10_10
5	x	x	x	x	x	x
8	x	x	x		x	x
9	x	x	x	x	x	x
10	x	xx	x	x	xx	xx
11		x		x	x	x
12		x				x
15		x			x	x
16		x			x	x
19	x	x	x	x	x	x

Table 16 Total network clearance time for all cost configurations for the coupled scheme

Combination	Sink costs		
	-1	$-t$	$t - T$
0_0	98	98	98
0_5	72	62	62
0_10	54	54	54
5_0	72	67	67
5_5	57	52	62
5_10	48	45	48
10_0	67	67	67
10_5	53	48	48
10_10	48	48	48

Table 17 Danger zone
clearance time for all cost
configurations for the coupled
scheme

| | Sink costs | | |
Combination	-1	$-t$	$t-T$
0_0	56	56	56
0_5	39	39	39
0_10	33	31	33
5_0	56	56	56
5_5	31	31	39
5_10	25	31	31
10_0	56	56	56
10_5	31	31	31
10_10	31	31	31

Table 18 Arcs selected to be
reversed in the evacuation
trees for cost configuration
-1

| Arcs | Contraflows allowed | | | |
Selected	0_5_-1	0_10_-1	0_15_-1	20_0_-1
(424, 425)	x	x	x	x
(426, 441)	x	x	x	x
(431, 432)			x	
(434, 435)			x	
(437, 556)		x	x	x
(438, 535)	x	x	x	x
(439, 438)		x		x
(440, 439)		x	x	x
(441, 440)		x	x	x
(443, 897)				x
(489, 631)				x
(507, 646)				x
(513, 902)	x			
(526, 527)				x
(528, 526)				x
(552, 550)				x
(582, 660)		x	x	
(587, 400)			x	x
(617, 599)				x
(660, 902)		x		
(711, 713)			x	x
(769, 771)			x	x
(773, 775)			x	
(775, 589)			x	
(777, 778)	x	x	x	x

Table 19 Arcs selected to be reversed in the evacuation trees for cost configuration $-t$

Arcs Selected	Contraflows allowed			
	0_5_-t	0_10_-t	0_15_-t	0_20_-t
(424, 425)	x	x	x	x
(426, 441)	x	x	x	x
(427, 428)		x	x	x
(431, 432)		x	x	x
(432, 433)				x
(433, 434)				x
(434, 435)		x	x	x
(437, 556)				x
(438, 535)		x	x	x
(443, 897)				x
(528, 526)		x	x	x
(541, 902)		x	x	x
(587, 400)			x	x
(641, 648)				x
(769, 771)				x
(777, 778)	x	x	x	x
(777, 779)		x	x	x

Table 20 Arcs selected to be reversed in the evacuation trees for cost configuration $t - T$

Arcs Selected	Contraflows allowed			
	0_5_3	0_10_3	0_15_3	0_20_3
(424, 425)	x	x	x	x
(425, 426)				x
(426, 441)	x	x	x	x
(437, 556)	x	x	x	x
(438, 535)	x	x	x	x
(439, 438)		x	x	x
(441, 440)		x	x	x
(440, 439)		x	x	x
(507, 508)				x
(528, 526)			x	x
(541, 902)			x	x
(553, 550)				x
(587, 400)		x	x	x
(594, 596)			x	x
(596, 612)			x	x
(711, 713)			x	x
(768, 584)				x
(769, 771)		x	x	x
(777, 778)	x	x	x	x
(777, 779)				x

Table 21 Nodes selected to be diverged for cost configuration -1

Nodes Selected	Divergence allowed			
	5_0_-1	10_0_-1	15_0_-1	20_0_-1
398	x	x	x	x
427			x	x
428	x	x	x	x
436	x	x	x	x
535	x	x	x	x
550				x
584		x	x	x
589			x	x
599			x	x
608		x	x	x
615			x	x
622		x	x	x
648		x	x	x
718				x
902		x	x	x
903		x		

Table 22 Nodes selected to be diverged for cost configuration $-t$

Nodes Selected	Divergence allowed			
	5_0_-t	10_0_-t	15_0_-t	20_0_-t
398	x	x	x	
401				x
428		x	x	x
436	x	x	x	x
437			x	
441			x	
535	x	x	x	x
564				x
584			x	x
589			x	
599			x	x
604				x
608		x	x	x
615			x	
622		x	x	x
631				x
643				x
648		x	x	
686				x
696				x
718				x
902	x	x	x	x

Table 23 Nodes selected to be diverged for cost configuration $t - T$

Nodes Selected	Divergence allowed			
	5_0_3	10_0_3	15_0_3	20_0_3
392				x
393		x		
398	x	x	x	x
413				x
427			x	x
428	x	x	x	x
436	x	x	x	x
437				x
440			x	x
441			x	
535	x	x	x	x
550				x
584		x	x	x
589		x	x	x
596				x
599			x	x
608			x	x
614				x
615		x	x	x
622		x	x	x
648				x
716			x	
779	x			
902		x	x	x

Table 24 Nodes selected to be diverged in the evacuation trees of the coupled scheme for cost configuration −1

Nodes Selected	Divergence allowed															
	5_5	5_10	5_15	5_20	10_5	10_10	10_15	10_20	15_5	15_10	15_15	15_20	20_5	20_10	20_15	20_20
394															x	
400	x				x			x	x	x	x		x	x	x	x
401												x				
409											x			x		
415													x			
427												x				x
428	x	x	x	x	x	x	x	x	x	x	x	x	x	x	x	x
436	x	x	x		x	x			x	x		x	x	x		x
477														x		
488														x		
491												x				
504															x	
512						x										
527						x	x		x	x	x	x	x		x	x
550							x			x	x	x	x		x	x
551											x					
552							x									
556		x	x		x	x	x	x	x	x	x		x	x	x	x
560													x	x	x	x
562								x								
564												x		x		
571												x			x	
572															x	x
575														x		
577											x		x		x	

	584	591	596	599	604	607	608	609	610	613	615	616	620	622	627	631	646	648	686	694	696	702	706	716	771	871	902	903
	X		X		X					X	X	X	X	X						X		X						
	X												X				X	X		X			X					
	X			X			X					X	X													X	X	X
	X	X					X						X	X				X			X				X		X	
	X												X					X								X	X	
	X									X			X	X			X	X										
	X	X					X						X			X	X	X									X	X
	X		X				X			X			X					X									X	X
	X		X						X			X	X															
								X					X					X								X	X	X
												X	X														X	X
										X			X		X			X									X	
	X																											
																		X										
																		X									X	
													X														X	

Table 25 Arcs selected to be reversed in the evacuation trees of the coupled scheme for cost configuration −1

Arcs	Contraflows allowed															
Selected	5_5	5_10	5_15	5_20	10_5	10_10	10_15	10_20	15_5	15_10	15_15	15_20	20_5	20_10	20_15	20_20
(424, 425)	×	×	×	×	×	×	×	×	×	×	×	×		×	×	×
(426, 441)	×	×	×	×	×	×	×	×	×	×	×	×	×	×	×	×
(427, 428)				×			×	×			×	×			×	×
(428, 431)				×				×				×	×			×
(431, 432)		×	×	×	×	×	×	×		×	×	×	×	×	×	×
(432, 433)				×			×	×				×				×
(433, 434)				×			×	×				×				×
(434, 435)		×	×	×		×	×	×		×	×	×		×	×	×
(438, 535)		×	×	×	×	×	×	×		×	×	×	×	×	×	×
(440, 439)	×			×			×	×				×				×
(441, 440)	×			×			×	×				×				×
(443, 897)			×	×				×			×	×			×	×
(528, 526)		×	×	×			×	×			×	×			×	×
(541, 902)		×	×	×				×			×	×			×	×
(587, 400)		×	×	×			×	×			×	×			×	×
(711, 713)		×	×	×			×	×			×	×			×	×
(769, 771)		×	×		×	×	×	×	×		×	×	×		×	×
(777, 778)	×	×	×	×	×	×	×	×	×	×	×	×	×	×	×	×
(777, 779)		×	×	×		×		×	×	×	×	×		×	×	×

Table 26 Nodes selected to be diverged in the evacuation trees of the coupled scheme for cost configuration −t

Nodes Selected	Divergence allowed															
	5_5	5_10	5_15	5_20	10_5	10_10	10_15	10_20	15_5	15_10	15_15	15_20	20_5	20_10	20_15	20_20
392			x								x			x		x
394										x	x				x	x
400					x	x	x	x	x		x	x	x		x	
401												x	x			
409			x							x		x		x		x
410													x			
413										x						
414																x
415														x		
427											x	x				x
428	x	x		x	x	x	x	x	x		x	x	x	x	x	x
431								x				x				x
435										x						
436	x	x			x			x	x				x	x		x
504																x
527						x				x	x				x	
537										x						
550											x					
551			x													
552					x	x	x	x	x							
556										xx	x	x		xx	x	
560												x		x		x
562			x											x		x
564														x		x

(continued)

Table 26 (continued)

Nodes Selected	Divergence allowed															
	5_5	5_10	5_15	5_20	10_5	10_10	10_15	10_20	15_5	15_10	15_15	15_20	20_5	20_10	20_15	20_20
575															x	
576															x	
577													x			
584		x		x	x	x	x	x	x	x	x		x		x	
596											x	x				x
599											x	x				
604						x			x		x			x	x	x
609														x		
610									x	x		x		x		x
613										x						x
615			x				x	x			x	x				x
616												x	x			
619									x				x			
620					x	x	x	x	x	x	x		x			
622	x									x	x		x		x	
631													x		x	
636													x			
648	x									x						

	650	686	691	571	572	574	694	696	702	706	716	718	725	731	871	902	903
						×	×										
		×	×	×		×					×	×			×		
		×	×				×	×					×	×		×	
	×	×			×					×						×	×
										×							
		×				×											×
							×			×							
		×										×				×	
																	×
		×						×									
		×															
		×														×	
																×	

Table 27 Arcs selected to be reversed in the evacuation trees of the coupled scheme for cost configuration $-t$

Arcs Selected	Contraflows allowed															
	5_5	5_10	5_15	5_20	10_5	10_10	10_15	10_20	15_5	15_10	15_15	15_20	20_5	20_10	20_15	20_20
(424, 425)	×	×	×	×	×	×	×	×	×	×	×		×	×	×	×
(425, 426)			×							×		×				
(426, 441)	×	×	×	×	×	×	×	×	×		×		×		×	
(427, 428)			×	×			×	×			×	×			×	
(428, 431)														×		×
(431, 432)	×	×		×	×	×	×	×	×	×	×		×		×	×
(432, 433)				×			×	×		×	×	×		×	×	×
(433, 434)			×	×			×	×			×				×	×
(434, 435)		×	×	×		×	×	×			×	×			×	×
(435, 436)			×									×				
(438, 535)	×	×	×	×	×	×	×	×	×	×	×		×		×	
(439, 438)				×				×		×	×	×		×		
(440, 439)				×				×		×				×		×
(441, 440)			×	×				×		×		×		×		×
(443, 897)				×				×								
(493, 497)												×				×
(494, 493)																×
(498, 533)												×				×
(507, 646)																×
(526, 541)												×				
(527, 543)			×									×				

	1	2	3	4	5	6	7	8	9	10	11	12	13	14	15	16	17
(528, 526)		×			×	×			×	×	×		×		×		×
(531, 529)	×				×												
(532, 531)	×																
(533, 532)	×																
(540, 622)	×																
(541, 902)		×	×			×			×	×	×		×		×		×
(587, 400)		×				×			×	×			×				×
(641, 648)					×				×				×	×			
(711, 713)					×							×		×			
(722, 724)					×												
(768, 584)				×	×		×	×									
(769, 771)	×	×	×		×	×	×		×	×	×		×	×			×
(777, 778)	×	×	×		×	×			×	×	×		×	×	×	×	×
(777, 779)	×	×			×	×			×	×			×	×	×		×

Table 28 Nodes selected to be diverged in the evacuation trees of the coupled scheme for cost configuration $t - T$

Nodes Selected	Divergence allowed															
	5_5	5_10	5_15	5_20	10_5	10_10	10_15	10_20	15_5	15_10	15_15	15_20	20_5	20_10	20_15	20_20
392					x					x			x	x		
393						x								x		
394						x									x	x
400	x		x		x	x	x		x	x	x	x	x	x	x	x
409											x		x	x	x	x
413											x					
427							x		x	x		x	x	x	x	x
428	x	x	x	x	x	x	x	x	x	x	x	x	x	x	x	x
431															x	
435									x				x			
436	x	x	x	x	x	x	x	x	x	x	x	x	x	x	x	x
437					x	x			x				x			
527					x		x	x	x	x	x	x	x	x	x	x
550				x				x		x	x	x	x	x	x	x
551									x				x	x		
556			x	x			x	x	x	x	x	x	x	x	x	x
562														x	x	
584					x		x	x	x	x	x	x	x	x	x	x
593												x		x		
599													x		x	
608														x		x

	610	613	614	615	622	631	648	716	871	902	903
	×	×		×	×		×	×	×	×	×
	×	×					×	×	×	×	×
	×	×		×	×		×		×	×	
		×	×	×	×		×		×		
	×				×		×	×		×	×
	×	×		×	×		×		×		
		×		×	×	×	×		×		
			×	×	×			×	×		
	×				×		×		×		
	×						×		×		
		×		×	×		×		×		
				×	×				×		
	×								×		
									×		
				×	×				×		
					×				×		

Table 29 Arcs selected to be reversed in the evacuation trees of the coupled scheme for cost configuration $t - T$

Arcs Selected	Contraflows allowed															
	5_5	5_10	5_15	5_20	10_5	10_10	10_15	10_20	15_5	15_10	15_15	15_20	20_5	20_10	20_15	20_20
(424, 425)	×	×	×	×	×	×	×	×	×	×	×	×	×	×	×	×
(425, 426)								×				×				×
(426, 441)	×	×	×	×	×	×	×	×	×	×	×	×	×	×	×	×
(427, 428)			×	×		×	×	×			×	×			×	×
(428, 431)								×				×				×
(431, 432)	×	×	×	×	×	×	×	×	×	×	×	×	×	×	×	×
(432, 433)			×	×		×	×	×			×	×			×	×
(433, 434)			×	×		×	×	×			×	×			×	×
(434, 435)	×	×	×	×	×	×	×	×	×	×	×	×	×	×	×	×
(438, 535)		×	×	×		×	×	×	×	×	×	×	×	×	×	×
(439, 438)		×	×	×			×	×		×	×	×		×	×	×
(440, 439)		×	×	×			×	×		×	×	×		×	×	×
(441, 440)		×	×	×		×	×	×		×	×	×		×	×	×
(528, 526)				×		×	×	×			×	×				×
(541, 902)				×				×				×				×
(587, 400)				×				×				×				×
(641, 648)				×												
(711, 713)			×	×		×	×	×			×	×			×	×
(768, 584)				×												
(769, 771)			×	×				×		×		×		×		
(777, 778)	×	×	×	×	×	×	×	×	×	×	×	×	×	×	×	×
(777, 779)		×	×	×	×	×	×	×		×	×	×		×	×	×

Table 30 Total network clearance time for all cost configurations for the coupled scheme

| | Sink costs | | |
Combination	-1	$-t$	$t - T$
0_0	127	127	127
0_5	119	118	119
0_10	118	118	118
0_15	121	118	118
0_20	118	118	118
5_0	118	118	118
5_5	118	118	118
5_10	118	118	118
5_15	118	118	118
5_20	118	118	118
10_0	120	118	118
10_5	118	118	118
10_10	118	118	118
10_15	118	118	118
10_20	118	118	118
15_0	118	118	118
15_5	122	118	118
15_10	118	118	118
15_15	118	118	118
15_20	122	118	118
20_0	118	121	118
20_5	118	118	118
20_10	118	118	118
20_15	118	118	118
20_20	118	118	118

Table 31 Danger zone
clearance time for all cost
configurations for the coupled
scheme

	Sink costs		
Combinations	-1	$-t$	$t - T$
0_0	6	8	6
0_5	4	19	5
0_10	3	19	5
0_15	5	16	5
0_20	4	16	5
5_0	5	12	5
5_5	5	19	5
5_10	19	7	5
5_15	19	19	5
5_20	19	5	3
10_0	3	5	5
10_5	19	7	5
10_10	3	7	5
10_15	4	19	3
10_20	19	19	3
15_0	5	5	5
15_5	3	19	3
15_10	19	19	3
15_15	19	19	3
15_20	7	15	3
20_0	5	19	5
20_5	7	7	3
20_10	7	19	3
20_15	4	7	3
20_20	7	14	3

References

1. Andreas, A.K., Smith, J.C.: Decomposition algorithms for the design of a nonsimultaneous capacitated evacuation tree network. Networks **53**(2), 91–103 (2009)
2. Bayram, V.: Optimization models for large scale network evacuation planning and management: a literature review. Surv. Oper. Res. Manag. Sci. **21**(2), 63–84 (2016)
3. Beckmann, M.J., McGuire, C.B., Winsten, C.B.: Studies in the Economics of Transportation. Yale University, New Haven (1956)
4. Cova, T.J., Johnson, J.P.: A network flow model for lane-based evacuation routing. Transp. Res. A Policy Pract. **37**(7), 579–604 (2003)
5. Daganzo, C.F.: The cell transmission model: a dynamic representation of highway traffic consistent with the hydrodynamic theory. Transp. Res. B Methodol. **28**(4), 269–287 (1994)
6. Daganzo, C.F.: The cell transmission model, Part II: network traffic. Transp. Res. B Methodol. **29**(2), 79–93 (1995)
7. Department of Homeland Security: U S Department of Homeland Security, W.D., National response framework, 2013. https://www.fema.gov/media-library-data/20130726-1914-25045-1246/final_national_response_framework_20130501.pdf

8. Dosa, D.M., Grossman, N., Wetle, T., Mor, V.: To evacuate or not to evacuate: lessons learned from Louisiana nursing home administrators following Hurricanes Katrina and Rita. J. Am. Med. Dir. Assoc. **8**(3), 142–149 (2007)

9. Fleischer, L., Tardos, É.: Efficient continuous-time dynamic network flow algorithms. Oper. Res. Lett. **23**(3-5), 71–80 (1998)

10. Ford, L.R., Fulkerson, D.R.: Maximal flow through a network. Can. J. Math. **8**(3), 399–404 (1956)

11. Gale, D.: Transient flows in networks. Tech. rep., DTIC Document (1958)

12. Gurobi Optimization Inc.: Gurobi Optimizer Reference Manual (2016). http://www.gurobi.com

13. Hagberg, A., Swart, P., S Chult, D.: Exploring network structure, dynamics, and function using networkx. Tech. rep., Los Alamos National Lab. (LANL), Los Alamos (2008)

14. Hamacher, H., Tjandra, S.: Mathematical modelling of evacuation problems: a state of art. Tech. Rep. 24, Fraunhofer (ITWM) (2001). http://nbn-resolving.de/urn/resolver.pl?urn:nbn:de:hbz:386-kluedo-12873

15. He, X., Zheng, H., Peeta, S.: Model and a solution algorithm for the dynamic resource allocation problem for large-scale transportation network evacuation. Transp. Res. C Emerg. Technol. **59**, 233–247 (2015)

16. Hoppe, B., Tardos, É.: Polynomial time algorithms for some evacuation problems. In: SODA, vol. 94, pp. 433–441 (1994)

17. Hoppe, B., Tardos, É.: The quickest transshipment problem. Math. Oper. Res. **25**(1), 36–62 (2000)

18. IRFC: The Red Cross Red Crescent Approach To Disaster and Crisis Management (2011). http://www.ifrc.org

19. Kalafatas, G., Peeta, S.: Planning for evacuation: insights from an efficient network design model. J. Infrastruct. Syst. **15**(1), 21–30 (2009)

20. Kim, S., Shekhar, S.: Contraflow network reconfiguration for evacuation planning: a summary of results. In: Proceedings of the 13th Annual ACM International Workshop on Geographic Information Systems, pp. 250–259. ACM, New York (2005)

21. Kim, S., Shekhar, S., Min, M.: Contraflow transportation network reconfiguration for evacuation route planning. IEEE Trans. Knowl. Data Eng. **20**(8), 1115–1129 (2008)

22. Lu, Q., George, B., Shekhar, S.: Capacity constrained routing algorithms for evacuation planning: a summary of results. In: Advances in Spatial and Temporal Databases, pp. 291–307. Springer, Berlin (2005)

23. Marianov, V.: Location models for emergency service applications. In: Leading Developments from INFORMS Communities, pp. 237–262. INFORMS (2017)

24. Occupational Safety and Health Administration: Emergency preparedness and response: getting started. Evacuation & shelter-in-place. https://www.osha.gov/SLTC/emergencypreparedness/gettingstarted_evacuation.html. Accessed 1 May 2018

25. Peeta, S., Ziliaskopoulos, A.K.: Foundations of dynamic traffic assignment: the past, the present and the future. Netw. Spat. Econ. **1**(3–4), 233–265 (2001)

26. Pyakurel, U., Dhamala, T.N.: Continuous time dynamic contraflow models and algorithms. Adv. Oper. Res. **2016**, 7902460 (2016)

27. Pyakurel, U., Dhamala, T.N.: Continuous dynamic contraflow approach for evacuation planning. Ann. Oper. Res. **253**(1), 573–598 (2017)

28. Rebennack, S., Arulselvan, A., Elefteriadou, L., Pardalos, P.M.: Complexity analysis for maximum flow problems with arc reversals. J. Comb. Optim. **19**(2), 200–216 (2010)

29. Sheffi, Y.: Urban transportation networks, vol. 6. Prentice-Hall, Englewood Cliffs (1985)

30. Sheu, J.B., Pan, C.: A method for designing centralized emergency supply network to respond to large-scale natural disasters. Transp. Res. B Methodol. **67**, 284–305 (2014)

31. Shinouda, M.M.R.: Study of contraflow operations for hurricane evacuation. Ph.D. thesis, University of Alabama at Birmingham, Graduate School (2009)

32. Stepanov, A., Smith, J.M.: Multi-objective evacuation routing in transportation networks. Eur. J. Oper. Res. **198**(2), 435–446 (2009)

33. Transportation Networks for Research Core Team: Transportation networks for research. https://github.com/bstabler/TransportationNetworks. Accessed 1 Dec 2017
34. Transportation Research Board: The Role of Transit in Emergency Evacuation. Transportation Research Board Special Report 294 (2008). http://onlinepubs.trb.org/onlinepubs/sr/sr294.pdf
35. Tuydes, H., Ziliaskopoulos, A.: Tabu-based heuristic approach for optimization of network evacuation contraflow. Transp. Res. Rec. J. Transp. Res. Board (1964), 157–168 (2006)
36. Urbina, E., Wolshon, B.: National review of hurricane evacuation plans and policies: a comparison and contrast of state practices. Transp. Res. A Policy Pract. 37(3), 257–275 (2003)
37. Vogiatzis, C., Pardalos, P.M.: Evacuation modeling and betweenness centrality. In: International Conference on Dynamics of Disasters, pp. 345–359. Springer, Berlin (2016)
38. Vogiatzis, C., Walteros, J.L., Pardalos, P.M.: Evacuation through clustering techniques. In: Models, Algorithms, and Technologies for Network Analysis, pp. 185–198. Springer, Berlin (2013)
39. Wolshon, B.: "One-Way-Out": contraflow freeway operation for hurricane evacuation. Nat. Hazard. Rev. 2(3), 105–112 (2001)

Modelling Possible Oil Spills in the Barents Sea and their Consequences

Fuad Aleskerov and Sergey Demin

Abstract The oil and gas industry growth has increased rapidly in the Barents Sea during the last few years. The Arctic Zone is considered to be a relatively clean area. However, there is a certain number of "hot spots" in the Arctic due to the activities of extracting companies.

We studied the problems connected with the production of these two types of fossil fuel and carried out simulation model. This model shows the results of oil or gas flowing accident related to drilling complex, taking into account sea currents, winds, temperature, and ice. By using this model, we can highlight areas in the Barents Sea with the highest potential of the disaster so that preventive measures could be taken. In addition, this model helps to organize elimination of fossil fuel flowing consequences.

Introduction

The growth of the world population and the depletion of hydrocarbon resources increases attention paid to the reserves located in the remote regions, including offshore deep-water fields and fields located in Arctic waters. The enhancement of exploration and production increases the probability of a spill from offshore oil platforms and related pipelines, reservoirs for oil products storage, as well as from operations on shipment of oil [1].

At the same time, changing sea ice conditions opens new navigation routes in Arctic Zone [2]. The probability of oil spills increases with the number of ships and volumes of oil and its products, which are transported and used by vessels as fuel. It means denser vessel traffic over a longer period of navigation. However, a new sea route will create a navigable risks and related risks of oil spills.

F. Aleskerov (✉) · S. Demin (✉)
National Research University Higher School of Economics, Institute of Control Sciences of Russian Academy of Sciences, Moscow, Russia
e-mail: alesk@hse.ru; ssdemin@edu.hse.ru

© Springer Nature Switzerland AG 2018
I. S. Kotsireas et al. (eds.), *Dynamics of Disasters*, Springer Optimization and Its Applications 140, https://doi.org/10.1007/978-3-319-97442-2_2

In addition, we should take into account specific weather conditions in the Arctic Zone, such as lower temperature and salinity levels, formation and movement of pack ice, and long periods of the darkness (the polar night). Each of these factors might both increase the probability of fossil fuels spill and reduce the efficiency of pollution elimination process [3].

One more typical feature of Arctic Zone is fast ice, which has its own advantages and disadvantages. On the one hand, it interferes oil distribution in case of spill, which helps disaster consequences liquidation. However, on the other hand, in case of great amount of fuel it might get under the fast ice. If such happens, the elimination of all spilled oil becomes much harder [4].

And the last, but not the least characteristic of the Arctic Zone is specific types of animals. The overwhelming majority of local species have a relatively greater life expectancy and a slower cycle of alternation of generations. As a result, in case of accident, the restoration of flora and fauna after the disaster will be seriously slowed.

Given the fact that complete exclusion of fuel spillage is impossible, elimination measures planning is the most important factor for the improvement of works on liquidation of oil spills [1].

Methods of oil spills elimination are divided into three main categories. The first one is the mechanical recovery, where oil is retained in the area of the spill with the use of booms or natural traps and removed using skimmers and pumps. Second category, the nonmechanical recovery, uses chemicals to resist the spill. Incineration or biological treatment of oil pollution is used for decomposition or dispersion of the oil slick. Finally, the manual methods when oil is removed using conventional hand tools and techniques such as pails, shovels, or nets. And for better usage of all types of liquidation works, it is important to know approximate direction of oil flow, in order to choose the best way for resistance, paying attention to all factors, influencing the process efficiency.

Taking into account aforementioned facts, the problem of prediction of fuel spill trajectory in case of accident becomes crucial. Therefore, in this work we will construct a model, which will allow simulating oil spills trajectory. For this purpose, we will study the Barents Sea as an example.

Literature Review

Nowadays the main approach for fossil fuel spill simulation is the use of dynamic mathematical models. Unfortunately even in the most well-known models the presence of all components can hardly be found, allowing calculating the evolution of the oil slick on the water surface and land, and place them in contact in the coastal zone, but still adapted to Arctic conditions.

For instance, the model proposed by Pavlenko et al. [4] is based on continuity equation and Darcy's law, which describes the fluid flow through a porous medium by providing with the relationship between the instantaneous discharge rate through a porous medium, the viscosity of the fluid, and the pressure drop over a given distance. In this work the authors study oil spill in the coastal zone of the Barents Sea. The obtained solution has a relative ease of calculation and has potential for the further improvement by considering arbitrary relief of the surface of the soil and the source function, which will allow assessing oil pollution in coastal zones for various scenarios of accidental spills. However, it meets a few real situations, because certain number of unaccounted physical parameters, such as air temperature, which influence the process of oil distribution, are not considered.

Another model for oil spill simulation is the Oil Spill Contingency and Response (OSCAR) model [5, 6]. It takes into account the ice coverage, currents, and wind data. For this purpose, it includes an oil weathering, an oil combat, chemical fates, and a 3-dimensional oil trajectory submodels. However, in these parts spreading is calculated according to differential models, which makes the process of simulation time and resource consuming [7].

One more model—SpillMod—was introduced by Ovsienko et al. [8, 9] especially for computing the propagation of oil spills in open water and in case of specified conditions, characterized by spatial distribution of ice with various concentration. This model is based on the system of vertically averaged Navier–Stokes equations. SpillMod can give the following information: the trajectory of the spill spread and its form any time, dynamics of spill hydrocarbon balance (evaporation or natural dispersion in water column), and dynamics of changes in average of spill thickness [10].

In turn, Stanovoy et al. [11] constructed the model, which helps to study the spill in case of high ice concentration. A three-dimensional model OilMARS calculates the transfer and transformation of oil pollution on the sea surface as a result of a long emergency of oil spills from stationary or moving sources. This model uses the fields of cohesion and ice drift obtained by hydrodynamic models running in operational and forecasting mode.

Accidental oil spill is represented in the model by a sequence of discrete spills of meals that periodically come from the source of pollution to the surface water. It is assumed that the cohesion of the ice cover affects the area. As a result, it increases oil slick thickness in the processes of spreading and further diffusion. In addition, OilMARS accounts for the following processes: spreading on the surface of the snow or on the ice surface without snow, vertical and horizontal absorption into snow and vertical absorption into ice (using a modified method based on Darcy's law), and drift with the ice field.

Nevertheless, this model also has some disadvantages. For instance, for the correct usage of model we must set many parameters of local and regional ice conditions, which sometimes become a real problem [12]. Then, one can use another model [13]. Among other factors, this model allows simulation of such processes as oil freezing in a solid ice body, oil trapping between ice fragments in broken ice fields.

Even in case of considering most of these factors like in [14], it is important to take into account how oil changes its physical characteristics and chemical compositions over time. These changes are considered to be weathering, which includes evaporation of the oil drop dispersion in water, emulsification of oils, and biodegradation. Certainly, all these factors influence oil remaining in the environment.

However, oil spill simulation model is not enough for prediction of the process. All models for the high quality result must be applied to correct and precise data. Unfortunately, it is usually hard to give accurate information considering the amount of fossil fuel spilt in the water. Therefore, in practice some templates of spillage are used.

For instance, Zatsepa et al. [15] use the following patterns for different situations according to the type of the accident:

1. oil transporting barge—50% of total capacity,
2. stationary and floating oil rig and oil terminals—1500 t,
3. pipeline break—25% of the maximum flow volume for 6 hours and the oil volume between the isolation valves on ruptured pipeline section,
4. pipeline puncture—2% of maximum volume pumped for 14 days,
5. fixed storage facilities of oil and oil products—100% of the volume of the maximum capacity of a single storage object.

Furthermore, it is important to point out that in addition to volume, time, and place oil spills are characterized by the intensity, dynamics of leakage, and the associated duration of the spillage. Consideration of these characteristics may have an impact on the estimated behavior of spills and choosing the best measures for their localization and liquidation.

Similarly, for the leakage volume some patterns are used. Sidnyaev et al. [16] mentioned the following examples:

- steady oil spillage from the emergency tanker with the duration of approximately 10 hours,
- 50% of the spill gets into the sea immediately, and the remaining 50% over the next 24 hours,
- 25% of the volume of the damaged tanks, located above the holes, falls into the water within 20 minutes, the rest is displaced within the next 24 hours,
- 25% of the leakage gets into the sea during the first hour, the remainder falls in the sea within the next 12 hours.

Beyond that, even, if oil spills' time and place are obtained or estimated, one should take into account air flows, which also influence spillage movement. Lesov [17] for this purpose proposed to use the model based on a system of Euler equations under the assumption of perfect fluid dynamics.

As it was mentioned, no dynamic mathematical model accounts for all components, allowing calculating the evolution of the oil slick on the water surface. In addition, it also requires large computing resources. For this purpose it is important to note the model of oil spill simulation by networks.

Zhilyakova [18] considers sea as a network of small regions, connected by edges to the four main directions (North, East, South, and West). In turn, the oil spill is considered like a resource, which can be stored in nodes of the network. As a result, iterative model is constructed, where on each step every node with nonzero amount of the resource transfers it to all of its neighbors. Transferring volume of the resource is calculated for each direction by formula considering the direction and power of air and water flows.

A Model

In our model, we used the approach proposed in [19] for modelling of air pollution propagation. According to this approach, the amount of pollutant emission transferred from certain elementary area is proportional to the amount of polluted area in this area, wind intensity, and the difference of heights.

In our model the list of the parameters, characterizing the pollution propagation, differs from the model from [19]. This model uses the diffusion equation with decomposition on different environment. However, the main idea of construction of oil distribution function remains the same. The main difference is the weighted linear combination of the components (wind and flow speed—V_{wind}, V_{flow} accordingly) in comparison with multiplication in the previous model, i.e.,

$$Q_{distiruted} = Q_0 e^{a+bV_{flow}+cV_{wind}}, \qquad a, b, c \in R \qquad (1)$$

It is important to understand that in certain sea area there is the main flow with its own power and direction. However, oil distribution occurs in other directions as well. In order to consider this fact, we introduce V_{flow} and V_{wind}—velocities of water flow and wind in other directions.

However, temperature also influences the rate of oil distribution. It is explained by the fact that in case of low temperature all processes proceed slower. For this reason we propose to use increasing function $a(T)$ instead of certain coefficient a, which will show the distribution rate dependence on ambient temperature. This function should have complex structure, because in some value intervals temperature may influence linearly, while in some other value intervals dependence might be much more sophisticated. As a result, we got the following formula for the amount of distributed oil (formula 2):

$$Q_{distiruted} = Q_0 e^{a(T)+bV_{flow}+cV_{wind}}, \qquad b, c \in R \qquad (2)$$

One additional component of our model arises from the presence of ice in the area under study. For this purpose, we impose the condition which blocks pollution propagation through the ice (formula 3). Due to this condition, the whole amount of oil on the territory with ice is not considered for the distribution. By this setting, we take into account the oil transferring interference by the ice

$$Q_{distiruted} = \begin{cases} Q_0 e^{a(T)+bV_{flow}+cV_{wind}}, & no\ ice \\ 0, & elementary\ territory\ with\ ice \end{cases} \qquad b, c \in R \qquad (3)$$

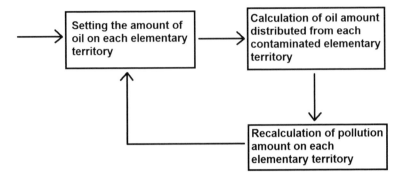

Fig. 1 Scheme of the simulation algorithm

After defining oil propagation function, we start the simulation process. For this purpose, on each iteration we calculate for each elementary territory with oil the amount of transferred pollution, according to formula (4).

$$\widehat{Q}_{ijk} = Q_{(i-1)j} e^{a(T_j)+bV_{flow,j}+cV_{wind,j}}, \qquad b, c \in R \qquad (4)$$

Here $\widehat{Q}_{ijk}$ is the amount of transferred oil from point j to point k at the iteration i, $Q_{(i-1)j}$ is the amount of pollution on territory j at the iteration $(i-1)$, T_j is the temperature on the territory j, and $V_{flow,j}$ and $V_{wind,j}$ are velocities of sea current and wind on the territory j.

Afterwards, we sum the pollution from all territories (formula 5) and obtain new amount of oil in the contaminated territory (Figure 1).

$$Q_{ik} = Q_{(i-1)k} + \sum_{j=1}^{N} \widehat{Q}_{ijk} - \sum_{j=1}^{N} \widehat{Q}_{ikj}, \qquad \forall k = 1, \ldots, N \qquad (5)$$

Application of the Model

For application of our model we have chosen the Barents Sea. We divided it into elementary territories about 2 km^2 each. Obviously, the square of this elementary territory might be taken much smaller. For all calculations we used winds and flows maps obtained from the Hydrometeorological Research Centre of Russian Federation (Hydrometcentre of the Russia) [20]. It is important to note that if we consider all Barents Sea surface due to the ratio of the sea surface area to the mean area polluted by the oil spillage the simulation of pollution propagation can be calculated only roughly. In order to get more detailed prediction we should focus

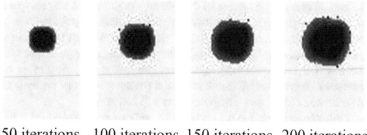

50 iterations   100 iterations   150 iterations   200 iterations

Fig. 2 Application of the model in the area without strong flows

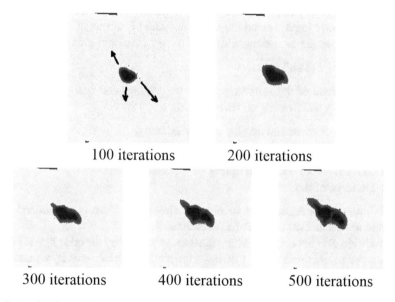

100 iterations   200 iterations

300 iterations   400 iterations   500 iterations

Fig. 3 Application of the model in the area with flows. On the first scheme arrows show the direction of sea currents

on a certain part of the Barents Sea. In addition, obtaining more precise data will help us to refine the simulation process.

However, even the data set used allows to achieve certain results, which can be very useful for propagation prediction. In Figures 2 and 3 two examples of application of our model for certain parts of Barents Sea are presented, which show the process of the oil propagation in sequential periods of time.

In these two figures the model highlighted the area with oil. For clearer and more precise understanding of the situation, we propose gradual coloring of the polluted area depending on the quantity of oil: dark black color means large amount of oil, while gray color—a small one. The threshold values between different colors are the parameters of the model, which can be specified before the simulation, i.e., we allow the user to choose the system of the threshold values with respect to a certain situation under study. The reason is that sometimes in case of polluted area surface assessment for evaluation of environmental damage even small amount of oil can be dangerous for fish or other sea inhabitants. Therefore, an exponential system of the thresholds is necessary. Meanwhile, in case of assessment the surface for spill liquidation small amount of oil is not so crucial (because it can be easily removed). Thus, in some other situations a linear system will be better.

We did not present here the dependence of spillage propagation on the surface temperature. It is taken into account by selection of corresponding function a in the formula (1).

As it was mentioned in the literature review, in case of no information about the type of the oil spillage experts usually simulate certain number of standard scenarios. We have implemented the same approach in our model. For this purpose, we introduce several parameters, which can be separated into groups:

1. The location of the oil spillage.

   (a) Coordinates of the starting point of the spillage (oil transporting barge or stationary and floating oil rig location),

2. The amount of oil leaked into the sea immediately,
3. A regime of the consequent spillage.

   (a) The amount of leaked oil during the period of time,
   (b) Duration of the spillage.

All these parameters allow us to simulate almost any situation mentioned in the literature review and used in real-life situations.

In addition, it is important to assess the damage caused by the oil spillage. For this purpose, our model on each step calculates the polluted surface area by summarizing area of all contaminated territories with special coefficient, which shows the degree of contamination. We propose this coefficient in order to take into account the size of our elementary territories. In addition, it helps to consider the fact that covering of the surface does not occur immediately. It means that in case of containing by an elementary section small amount of oil (for example, 0.1 t) it is not fully covered by oil.

Moreover, propagation of the fossil fuels in the sea can cause damage to the local fauna. For evaluation of this type of damage, we use the map of fishing territories in the Barents Sea (http://projects.scanex.ru/RussianArcticMSP-Barents) and calculated the contaminated surface area within fishing territories. Certainly, we understood the fact that this assessment will be approximate because of many facts (fishing territories are highlighted roughly, there might be other types of fish, which are prohibited for fishing and, as a result, are not taken into account). However, this feature of our model can be improved in case the required data is obtained.

Conclusion

We constructed the model, which helps to simulate the propagation of the oil spillage, calculate the contaminated surface area and the part of polluted area within fishing territories in the Barents Sea. The latter will help to assess the damage of the oil spillage to the local fauna. Important feature of the model is that the polluted area can be evaluated very fast allowing to organize efficiently the process of cleaning the contaminated area.

The parameters of the model can be specified by the user for her own tasks, such as system of threshold values for contaminated area or the scenario of the pollution.

Acknowledgements The model was developed within the framework of the Basic Research Program at the National Research University Higher School of Economics (HSE) and supported within the framework of a subsidy by the Russian Academic Excellence Project "5-100".

We are grateful for the comments and suggests of the anonymous reviewers.

References

1. WWF Report: Oil Spills: Problems Connected with Liquidation of Consequences of Oil Spills in Arctic Seas (in Russian). 2nd edn, augmented (2011)
2. Hassol, S.: Impacts of a Warming Arctic. Arctic Climate Impact Assessment. AMAP, CAFF, and IASC. Cambridge University Press, New York (2004)
3. Pavlenko, V.: The environmental standard for the Barents (in Russian). The oil of Russia, No. 4, pp. 71–76 (2012)
4. Pavlenko, V.I., Muangu, J., Korobov, V.B., Lokhov, A.S.: Actual problems of prevention and liquidation of oil spills in the Arctic and methods of environmental damage to coastal areas assessment (in Russian). Arctic Ecol. Econ. **3**(19), 4–11 (2015)
5. Reed, M., Aamo, O.M., Dling, P.S.: Quantitative analysis of alternate oil spill responce strategies using OSCAR. Spill Sci. Tech. Bull. **2**, 67–74 (1995)
6. Reed, M., Singsaas, I., Daling, P.S., Faksness, L.G., Brakstad, O.G., Hetland, B.A., Hokstad, J.N.: Proceedings of the 2001 Oil Spill Conference. Modeling the Water Accommodated Fraction in OSCAR2000, Tampa, Florida (2001)
7. Mackay, D., Buist, I., Mascarenhas, R., Paterson, S.: Oil Spill Processes and Models, p. 94. R&D Division, Environmental Protection Service, Ottawa (1980)
8. Ovsienko, S., Zatsepa, S., Ivchenko, A.: 15th International Conference on Port and Ocean Engineering under Arctic Conditions. Study and Modeling of Behavior and Spreading of Oil in Cold Water and in Ice Conditions, Espoo, Finland (1999)
9. Ovsienko, S.N., Zatsepa, S.N., Ivchenko, A.A.: Oil Spill Modeling and Environmental Risk Assessment (in Russian). Hydrometeoizdat, No. 209 (2005)
10. Ovsienko, S.N., Zatsepa, S.N., Ivchenko, A.A.: Numerical Simulation as Data Support Feature for Decision Making when Selecting Strategy for Marine Environment Protection from Oil Pollution (in Russian). Hydrometeoizdat, No. 213 (2011)
11. Stanovoy, V.V., Lavrenov, I.V., Neelov, I.A: Oil Spill Simulation System for the Arctic Seas. Problems of the Arctic and Antarctic, No. 77 (2007)
12. Bronson, M., Thompson, E., McAdams, F., McHale, J.: Proceedings 25th Arctic Marine Oil Spill Program Technical Seminar (AMOP). Ice Effects on Barge Based Oil Spill Response Systems in the Alaska Beaufort Sea, Calgary, Alberta, Canada (2002)

13. Drozdowski, A., Nudds, S., Hannah, C.G., Niu, H., Peterson, I.K., Perrie, W.A.: Review of Oil Spill Trajectory Modelling in the Presence of Ice. Canadian Technical Report of Hydrography and Ocean Sciences (2011)
14. Westway Expansion Project Oil Spill Modeling, Portland, Oregon (2015)
15. Zatsepa, S.N., Solbakov, V.V., Stanovoy, V.V.: The experience of creating operational models for calculating the distribution of oil spills in the Barents Sea (in Russian). Arctic Ecol. Econ. **4**(16), 68–76 (2014)
16. Sidnyaev, N.I., Kuzmina, M.S., Metsherin, I.V.: Evaluation of scenarios of oil spills in the offshore zone of the Arctic seas using the models of geological information procession (in Russian). Eng. Surv. **4**, 68–80 (2014)
17. Lesov, V.M: Regional hydrodynamic model of the Hydrometcentre of the Russian Federation (in Russian). 80 years of the Hydrometcentre of the Russian Federation, pp. 36–58 (2010)
18. Zhilyakova, L.Y.: The application of resource networks for modeling of distribution of substances in the aquatic environment (in Russian), vol. 2, pp. 46–51 (2011)
19. Aleskerov, F.: Taxation for improving regional ecological situation. Ecological Economics of Sustainability Conference, Washington DC, World Bank (1990)
20. The Hydrometeorological Research Centre of Russian Federation. http://193.7.160.230/web/esimo/barenc/hfcst/hfcst_barn.php

Prospects and Bottlenecks of Reciprocal Partnerships Between the Private and Humanitarian Sectors in Cash Transfer Programming for Humanitarian Response

Ioanna Falagara Sigala and Toyasaki Fuminori

Abstract As an alternative to commodity-based programming (in-kind aid), Cash Transfer Programming is attracting both humanitarian organizations' and institutional donors' attention. Unlike in-kind aid, Cash Transfer Programming transfers purchasing power directly to beneficiaries in the form of currency or vouchers for them to obtain goods and/or services directly from the local market. In distributing currency to beneficiaries, the private sector, especially financial service providers, plays a prominent role, due to the humanitarian sector's limited relevant resources. The present work unveils challenges for the private and humanitarian sectors, which hinder implementing Cash Transfer Programming. Based on primary and secondary qualitative data, the paper presents the main characteristics and the mechanisms of Cash Transfer Programming to explore how the private sector is involved with Cash Transfer Programming. Then, this study presents bottlenecks of reciprocal relationships between financial service providers and humanitarian organizations in Cash Transfer Programming.

Introduction

Conducting efficient, effective, and fair humanitarian operations is an earnest desire of humanitarian organizations (HOs). For the attainment of this goal, Cash Transfer Programming (CTP) has attracted HOs' increasing attention as an alternative to in-kind assistance (commodity-based programming) [1, 2]. For the delivery of both,

---

I. Falagara Sigala (✉)
Research Institute for Supply Chain Management, WU (Vienna University of Economics and Business), Vienna, Austria
e-mail: ioannna.falagara.sigala@wu.ac.ata

T. Fuminori (✉)
School of Administrative Studies, York University, Toronto, ON, Canada
e-mail: toyasaki@yorku.ca

© Springer Nature Switzerland AG 2018
I. S. Kotsireas et al. (eds.), *Dynamics of Disasters*, Springer Optimization and Its Applications 140, https://doi.org/10.1007/978-3-319-97442-2_3

57

cash and in-kind aids, HOs are entering into partnerships with the private sector as a means to address the complex humanitarian problems that exceed the ability of a single organization and sector and to offer better services and products to beneficiaries [3, 4].

CTP aims to transfer purchasing power directly to households and individuals in the form of cash or vouchers for them to obtain goods and/or services directly from the local market. CTP allows beneficiaries to access health services, food, transportation, and education by providing them with cash, vouchers, or electronic transfers [5, 6]. Despite the increasing use of CTP in humanitarian aid, it accounted for only 10% of humanitarian assistance in 2016 [7]. To boost the share of CTP-based aid, the humanitarian sector and major institutional donors launch international and comprehensive efforts [8]. For example, via the Grand Bargain signed at the World Humanitarian Summit 2016 in Istanbul, Turkey, major donors and HOs made a public commitment to increase the share of CTP in their humanitarian responses [9].

Previous relevant research [1, 6, 10–13] on CTP concludes that aid provided via CTP allows HOs to achieve more cost-efficient and effective aid than in-kind aid because CTP enables HOs to employ more straightforward logistics. However, HOs' limited relevant resources and technologies to transfer purchasing power to beneficiaries cause a bottleneck in promoting CTP. Thus, the participation of the private sector, especially of financial service providers (FSPs), is essential to execute and have CTP prevail [14, 15]. FSPs are entities that provide financial services, including e-transfer services. Depending on the context, FSPs include e-voucher companies, financial institutions (such as banks and microfinance institutions) [8].

Private–humanitarian partnerships are assumed to have positive results for both sides and especially for HOs since private companies could provide them with technical expertise, and innovative solutions [3, 16, 17]. Recent literature on private–humanitarian partnerships focuses on relationships between HOs and logistics service providers (LSPs). A research of Bealt et al. [18] explored barriers and benefits of establishing collaborative relationships between HOs and LSPs. Lack of trust between partners, poor governance, and accountability of HOs as well as lack of process and clear visibility are seen as the main barriers for building a successful partnership between HOs and LSPs. Cozzolino et al. [16] explore the contribution of logistics service providers' initiatives to disaster relief and how LSPs are engaged with the humanitarian sector. Outsourcing of humanitarian logistics to commercial LSPs is also investigated in Vega and Roussat [19]. They found that LSPs could play a significant role in relief response and they could take the role of coordinator, operator, and partner in the different disaster phases. For the implementation of cash aid, financial service providers appear to play an important role by providing technology and expertise in transferring money to beneficiaries. Despite the recent growth of humanitarian sector's expectation toward CTP, relevant literature is limited [20, 21].

A systematic literature review on CTP from Doocy and Tappis [20] concludes that only nine studies are in peer-reviewed publications and none of them focus on the role of the private sector. The majority of peer-reviewed articles are case studies about CTP implemented in different geographical areas while others modelling the cost efficiency and effectiveness of cash versus in kind. For example, applying an economy-wide modelling framework and a social accounting matrix to the long-term famine crises of Ethiopia in 2000s, Gelan [22] uses an economy-wide modelling approach to examine the effectiveness and efficiency of cash and in-kind aid. He found that cash aid provided efficiency gains from savings on logistics, avoided disincentives to local food production, and had greater multiplier effects. Margolies and Hoddinott [12] applied activity-based costing methods to interventions in four countries (Ecuador, Niger, Uganda, and Yemen) and they have found that the per-transfer cost of providing cash is always less than food. Based on a case study of the famine crisis in Somalia in 2011, Ali and Gelsdorf [1] also indicate that a large-scale cash program could be even more successful than the distribution of foods. Simpson et al. [23] investigate the parameters which can be measured to determine the added value of employing CTP on water and sanitation programs.

To the best of our knowledge, all previous relevant academic literature [1, 6, 11–13] is evidence-based research which focuses on the effectiveness of CTP versus in-kind aid. Another stream of modelling literature of CTPs investigates the relationship between cash and education and nutrition of beneficiaries in development aid. Combining modelling techniques and evidence data, researchers found that conditional cash is positively correlated with education and school enrollment and school attendance [24–26]. Also, previous research shows that cash transfers have demonstrated large impacts on health and nutrition of children [27, 28]. Exceptionally, Andersson et al. [14] address a bottleneck of CTP associated with HOs' limited abilities to implement CTP. The authors indicate that HOs participating in the political crisis of Bosnia in 1997 did not have the necessary skills and resources to implement CTP smoothly. Despite the fact that the private sector plays an essential role in the execution of CTP, academic research, which sheds light on this area, is limited. Recently, Heaslip et al. [10] explore the CTPs' impact on humanitarian logistics from the perspective of supply chain strategies. Based on empirical data, the authors conclude that developing strategic partnerships between commercial companies and aid agencies is a determinant of successful CTPs. However, they do not conduct an in-depth analysis on the issue, including obstacles to overcome for partnerships between the private and humanitarian sectors.

The present exploratory study focuses on the role of the private sector, specifically of FSPs in the delivery of CTP. It also analyzes the mechanism of CTP, which would expose benefits, risks, and challenges. We employ a qualitative approach accessing primary data from interviews with experts from HOs and the private sector and secondary data from HOs' and institutional donors' reports as well as from academic articles. We have interviewed cash experts in HOs (DRC, NRC, CRS, WFP, Red Cross, and Welthungerhilfe), DG ECHO of the European Commission, and from private sector MasterCard and the RedRose, which is a private technology company that offers software for the execution of CTPs. We have also talked to two independent consultants who are collaborating with HOs and with development agencies for CTP. In addition, Cash Learning Partnership (CaLP) and UNHCR

Table 1 List of the interviews

Organization	Country	Date
Danish Refugee Council (DRC)	Denmark	July 2017
Catholic Relief Services (CRS)	USA	July 2017
DG ECHO (European Commission)	Belgium	July 2017
Consultant of CTPs for HOs	France	July 2017
Consultant of CTPs for HOs	USA	July 2017
WFP	Italy	July 2017
Red Cross	Canada	September
Norwegian Refugee Council (NRC)	Norway	October 2017
Welthungerhilfe	Germany	November 2017
MasterCard	USA	September 2017
RedRose	UK	September 2017

provided us with useful documents for this research. The interviews are conducted from July 2017 to October 2017. Table 1 presents the organizations that we have interviewed.

The structure of the paper is as follows. Following the section of introduction, section "Overview on the Mechanism of CTP" overviews the mechanism of CTP. After overviewing CTP, in section "Advantages And Disadvantages Of CTP From Hos' And Beneficiaries' Perspectives," we discuss advantages and disadvantages associated with the implementation of CTP from the perspective of HOs and beneficiaries. Section "Private Sector Participation in CTPs" discusses private sector's motivation to participate in CTP and associated challenges. Lastly, section "Conclusions and Future Research" presents conclusions and future research avenues.

Overview on the Mechanism of CTP

As preliminary information to discussions in sections "Private Sector Participation in CTPs and Conclusions and Future Research," this section elaborates on the mechanism of CTP in terms of three aspects: relevant stakeholders in CTP, decision criteria that HOs apply to implement CTP, and delivery mechanism that HOs utilize to transfer purchasing power to beneficiaries. Looking into the mechanism of CTP allows us to recognize how the private sector is involved with CTP.

Stakeholders

Many indispensable stakeholders are involved with CTP. They are: HOs, donors (private and institutional), the private sector (FSPs, suppliers, logistics providers, wholesalers, and retailers), national government, and beneficiaries. The following explains each of the stakeholders in detail to obtain an insight on their roles and interactions.

HOs HOs are acting as implementing agencies for CTP. The role of HOs is to establish the mechanism for providing cash to beneficiaries, monitoring and evaluating the conditions, and managing the data collected from the field. HOs are also responsible for managing all contracts with financial and logistics providers for the distribution of cash as well as to coordinate the partnerships with other HOs and the private sector.

Donors Donors include institutional donors (e.g., European Commission Humanitarian Office (ECHO)) and private donors. HOs rely on donations from private and institutional donors to operate CTP. In terms of donated money to HOs, discrepancies between private and institutional donors can be observed. Unlike private donors, institutional donors usually execute their discretion over how HOs must use their donated funds, either for CTP or for in-kind programming. In many cases, their decisions are based on political criteria. In our interviews, a cash expert from CRS explained "some countries like US have recently decreased the amount of money given for CTP and they are funding more in-kind programs because they want to minimize the chances that pure cash will be used by potential terrorists." We have also found that institutional donors are willing to contract FSPs directly for CTPs, instead of via HOs. This new trend of fund flow from donors is discussed in detail in section "Private Sector Participation in CTPs."

Private Sector and FSPs The private sector includes financial and logistics service providers, suppliers of technologies, retailers and wholesalers in the affected areas. The private sector, especially FSPs, plays an essential role as a partner of HOs, in terms of planning, assessments, implementation, and preparedness for CTP. The private sector uses its own existing infrastructure, so that it can shorten implementation time and reduce costs, security risks as well as duplication of efforts [29]. Private provides services and products that HOs are either unable to provide or which they can provide in a better way [30]. The role of the private sector is discussed in detail in section "Private Sector Participation in CTPs."

National Governments National governments usually run large-scale cash programs and have the regulations in place to allow or reject CTP. In many cases, national governments have strong ownership and leadership of CTP in their countries. HOs are required to collaborate and use the existing infrastructure at the national level to achieve effectiveness and efficiency of CTP [29]. FSPs also have to follow national regulations with respect to data privacy and cash transfer in affected areas. A financial expert from MasterCard explained in our interview, "we always follow the laws that each country has in place with respect to cash transfers and we always sign agreements with organizations for data privacy."

Beneficiaries Beneficiaries may be defined at an individual or household level, and may be selected based on geography, age, wealth status, vulnerabilities, or other characteristics specific to the needs of the humanitarian basis. Depending on their family situations, HOs usually provide cash to a member of the family and not to each individual within the family [29].

HOs' Decision Criteria for the Implementation of CTP

This subsection introduces a typical HO's decision-making criterion applied when they determine the implementation of CTP.

HOs' decision-making processes consist of three major steps: market analysis, risk analysis, and selection of response options (i.e., how to deliver CTP). First, HOs assess whether the market or a neighboring market can supply goods; then, they examine accessibility to markets. Market assessment aims to confirm the impact of introducing CTP on the local market in an affected area. Market assessment also helps to distinguish between goods and services which can be purchased locally, or which require direct delivery or complementary support from HOs. HOs also look into interventions by governments and other humanitarian organizations, which would positively or negatively affect the total supply chain. Following the step of market analysis, HOs conduct risk analysis. If HOs recognize security issues, then they decide to use vouchers or e-cash as a distribution measure.

The next step is the selection of response options. In this step, HOs look into the necessity of providing beneficiaries with some tasks (i.e., either conditional or unconditional CTP). *Conditional cash* transfers require that recipients meet certain requirements before the transfer is fulfilled. Cash transfers with conditions are those given after the recipients have performed some tasks or activities as a condition of receiving the cash transfer [31]. *Unconditional cash* transfers are direct cash with no conditions or work requirements with the assumption that money will be used to meet basic needs or be invested in livelihoods [8]. Unconditional cash transfers are usually common in the aftermath of a disaster to cover the first needs of the affected populations. If the initial market assessment indicates that the local market does not have the capacity, then HOs either help local traders to build their capacities or provide in-kind aid from international markets.

At the same time, HOs examine the conditions and liquidity of the local financial systems. Our interviews reveal that liquidity issues can occur which result in delays on the delivery of cash to beneficiaries. In this case, HOs look for alternative solutions by physically importing cash of goods. In addition, HOs need to pay attention to the impact of currency exchange rates in the local black market in affected areas since some vendors only accept the unofficial exchange rate. Figure 1 presents the decision-making process which HOs apply in executing CTPs.

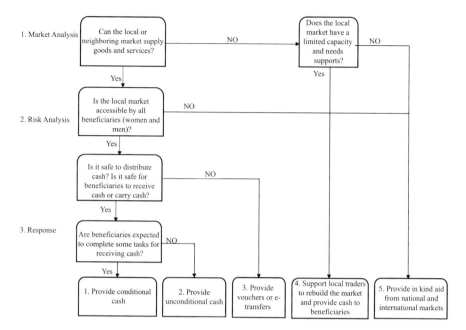

Fig. 1 Decision Tree of HOs for CTPs based on empirical findings

Delivery Mechanism

Once HOs determine to employ CTP, they next select the means of delivering cash or voucher transfer to beneficiaries. Based on primary and secondary data, this subsection explains the delivery mechanism of CTP.

Depending on conditions of the affected areas, CTP has the following four different means of delivering: *cash in hand, commodity vouchers, e-cash, and hawala transfer system.* Regardless of the modalities and delivery mechanism, cash is transferred to beneficiaries either conditionally or unconditionally, as explained in subsection "HOs' Decision Criteria for the Implementation of CTP."

Cash in Hand, in Envelopes, or Cheques Beneficiaries are provided with money directly from a HO or a third-party provider. Our interview with a cash expert from CRS confirmed that cash in hand is employed in "very remote areas, where there is no infrastructure and FSPs are not present. Also, we use them in the very first days of the response in order to give beneficiaries money to cover their first needs." Cash in hand requires that HOs have sufficient cash flow, staff members, and logistics support. Security in transporting and distributing cash and the acceptance by local authorities is essential. Cash in hand is cost-efficient because no transfer equipment and no literacy skills as well as training for beneficiaries are required. However, organizing logistics support for cash distribution to rural areas would generate costs. Additionally, during the process of cash distribution, corruption and/or security risks for staff members in charge of transportation need to be considered. Logistics providers which distribute cash and/or financial providers which issue checks are involved.

Cash or Commodity Vouchers Cash or commodity vouchers, for example, coupons, tokens, or smartcards, magnetic stripe cards, which can only be used in particular shops and/or on particular items from local markets are used from HOs. Private sector is involved with CTP as the supplier of coupons, tokens, or smartcards and as a retailer of items in case of commodity vouchers. Voucher is useful to collect data on beneficiaries' preferences. The data are used by HOs to avoid item shortages. As confirmed by an interviewed humanitarian expert, his organization has employed a private company to collect and analyze beneficiaries' consumption data. These data are then given to HOs to be used in future interventions: "our providers not only distribute cash to beneficiaries, but also monitor our programs and analyze data on beneficiaries' purchase preference." mentioned by a cash expert from Welthungerhilfe organization.

FSPs are also involved in cash or commodity vouchers, especially in using smartcards or magnetic stripe cards which need to link an agency bank account to an individual bank account. The use of smartcards or magnetic stripe cards requires connectivity and related infrastructure that can read and authenticate the cards such as PoS and ATMs, which generate additional costs for HOs. In the process of implementing, smartcards or magnetic stripe cards, HOs provide FSPs with beneficiaries' ID information. Although these modalities can speed the delivery of cash, they require training for the beneficiaries on how to use the cards, ATMs and on how to manage their PINs.

E-cash Recently, HOs are using *e-cash* for humanitarian interventions. E-cash is used in 34% of all CTPs around the world [8]. Many HOs are using mobiles to transfer cash. The system is based on SMS codes to beneficiaries' mobile phones which can be cashed out at a specific shop in affected areas. E-cash reduces total distribution costs for HOs as well as program recipients' costs of obtaining the cash transfer [32]. A cash expert from CRS elaborated that the use of e-cash is easier and more practical in developed countries and can support also the new phenomenon of refugees on the move: "... mobile transfers are increasingly used for the Syrian refugees especially in countries such Greece. For the beneficiaries on the move such as the refugees who moved from Turkey to Greece and from there to Germany, mobile and credit cards are the most appropriate modalities to use since they can use them in each country where they go." E-cash requires connectivity to a network, training to beneficiaries on how to use it, and the ID of beneficiaries. E-transfers could also involve biometric technologies, such as iris scanning technology to authenticate identities. E-cash has the highest level of preference among beneficiaries who either have previous familiarity and capacity to utilize technologies, such as phones [33]. FSPs, suppliers of advanced technologies, and retailers at the local markets are involved in e-transfers.

Hawala Transfer System   In some cases, HOs are using the so-called hawala transfer system. Hawala system is an informal money transfer system which is based on the trust that enables the transfer or remittance of money between two parties in a fast and inexpensive manner, without the direct involvement of a financial institution ([34], p. 327). This type of system was originally developed to facilitate trade between distant regions, where conventional banking institutions were either absent, incapable, or unsafe. In some cases, this system is used by HOs to distribute cash to areas where formal financial institutions are inaccessible or do not exist and in unsecure areas, for example, in Somalia and Syria.

A standard hawala transaction can be described as follows: a HO approaches a hawala dealer—called hawaladar—to request the transfer of money to another city or country or to a remote area. If they agree on the fees of services, the hawaladar contacts another hawaladar in the area of destination who organizes the delivery of money to the recipient. The beneficiary receives the money within a few hours since the remitting person ordered the transfer. As a result of this operation, an "outstanding debt between both hawaladars, which will have to be canceled out in the future, is generated." ([34], p. 328). It is actually a transfer of money without actual movement of money. A cash expert from NRC explained in our interviews: "this system provides the means to transfer money without a legal contract between the entry and the exit point. . . because the system is informal, we need an information system that can be used to track the money flow to beneficiaries and to report back to donors. Donors need a guarantee that their money is successfully delivered."

Advantages and Disadvantages of CTP from HOs' and Beneficiaries' Perspectives

To better understand HOs' motivations to use CTP, this section overviews advantages and disadvantages linked to CTP from the perspective of HOs and beneficiaries.

Advantages   Previous relevant references agree that CTP would respect beneficiaries' individual choice, which actually contributes to reducing the mismatch between beneficiaries' needs and HOs' aid [6, 35, 36]. Previous research [6, 13, 35, 36] indicates cost-efficiency of humanitarian aid as a major advantage of CTP. CTP is also expected to facilitate monitoring of its impact on households, markets, and communities. A cash expert from DRC explained ". . . CTP can provide more comprehensive feedback on people's needs, vulnerabilities, and coping strategies in addition to the humanitarian impact on local contexts and communities."

Disadvantages   Although CTP empowers recipients by allowing them to decide how the cash would be spent and it raises the bargaining power of women in the decision making [37], CTP may have inadvertently reinforced the traditional role of women of being responsible for daily household keeping [38]. CTP could also affect the power balance within household members and within community members, which sometime ignites a conflict between recipients and non-recipients. Defining targeted beneficiaries who enjoy aid by CTP is a sensitive task for HOs because the task often has inclusion and exclusion errors [14]. The errors would cause security risks for both beneficiaries and HOs: "beneficiaries who receive and organizations which deliver cash could be a target," said a cash expert from WFP.

HOs also need to pay attention to the negative impact of CTP on the local market, such as inflation caused by supply shortages and/or price control by oligopolistic or monopolistic firms. During the market assessment, HOs ensure whether the supply is reliable, prices are affordable, and markets will be able to respond to increases in demand which result from cash infusions without negative distortions [20].

Private Sector Participation in CTPs

HOs' partnership with FSPs is an essential element in delivering cash to beneficiaries [15]. This section highlights why FSPs are interested in participating in CTP and what their main opportunities are. This section also discusses bottlenecks for the further growth of reciprocal relationships between private and humanitarian sectors in CTP. Furthermore, based on our interview results, we discuss the new trend of donation flows from institutional donors to FSPs as well as its potential impact on HOs.

FSPs' Motivation for Participating in CTP

Three key drivers have been identified in relation to engagement of FSPs in CTP. As previous literature [15, 39] indicates, our interviews also confirm that FSPs' pursuance of corporate social responsibility (CSR) is one of their main motivations to engage with CTP. New business opportunities and the related profits are also FSPs major drivers to participate in CTP. FSPs recognize CTP as a fast-moving industry which allows them to enter into new markets and to develop new innovative technological products. Technology is changing the way HOs provide assistance, from experimenting with blockchain technology to provide cash to the use of biometrics to register and track beneficiary assistance through iris scans and fingerprinting [40]. Established companies such as Visa and MasterCard are working together with UN agencies and other HOs to develop new technological solutions in distributing CTPs and help HOs to scale up CTPs ([2], p. 25).

As the financial expert from MasterCard explained, "I think that there are multiple motivations behind the engagement of private sector in CTPs. The nature of crises is complex, and it needs a cross-sector thinking and analysis on how to do things differently . . . So, first is that we want to help, second there is an innovation angle. Taking a part in humanitarian operations gives us an opportunity to learn peoples' needs and to develop robust products and services. Third, participation in humanitarian response is a marketing tool for the company to attract young talents and also there is a commercial aspect and the sustainable growth both for the societies and companies."

Bottlenecks for Further Growth of Reciprocal Relationships Between FSPs and HOs

The analysis of our interview results leads to the conclusion that challenges related to collaboration between private and humanitarian sectors stem from the two sectors' perception gaps. Specifically, they are knowledge on finance, governance structure, market competition, and services offered, as well as humanitarian data protection.

Knowledge on Finance One of the challenges of the relationships between FSPs and HOs is the different expertise that may result in communication gaps and different perceptions. Bailey and Gordon [15] discuss that HOs are in a new territory, where they need to manage a significant amount of money and deliver it to thousands of beneficiaries. A humanitarian expert that provides consultancy to HOs confirmed, "HOs do not speak the financial language and are not familiar with national and international financial regulations. Actually, they do not know what they don't know." Humanitarian experts recognize that CTP could not be executed effectively and efficiently without FSPs: "we (HOs) really need FSPs' expertise, we cannot manage cash transfers without banks and the system behind . . . financial service providers are more effective and adapt quickly to our needs in some of the most difficulties places that we work" as a cash expert from NRC mentioned. HOs may need additional resources and more expert personnel to manage the relationships with FSPs.

Governance Structure and Planning Horizon Differences We learn from our interviews that the discrepancy of governance structures of FSPs and HOs hampers their collaboration in CTP. FSPs take centralized governance models, which allow them to execute faster decision making. In contrast, HOs usually take decentralized governance models, which enable the field to enjoy autonomy and flexibility in its decision making for rapidly changing affected areas' circumstances.

Budget planning discrepancies between the two sectors also block collaboration in CTP. The humanitarian sector applies a project-based and short-term budget planning. As a cash expert from Red Cross confirmed, "We usually get grants and donations after a disaster occurs, but we try to establish framework agreements with FSPs in different countries before disasters as part of our preparation plan." FSPs, on the other hand, usually work on long-term plans and need information for planning, such as information about the next few years' programs. HOs face difficulties to provide this kind of information, partly due to the nature of their work and their limited resources to work on forecasting and projections [15].

Market Competition and Service Offered HOs demand tailored solutions to adapt to the environment where they operate. This adaptation may not be affordable for FSPs in that the market is not large enough to pursue tailored services. FSPs are risk averse when it comes to tailored solutions: "we (FSPs) are thinking very much on scale basis which requires us to be a bit more conservative on what we offer, and we are not promising a lot in terms of customization. We explain what our technology does and if it does not fit the operations of our partners, we do not collaborate . . . " as confirmed by MasterCard expert.

Humanitarian Data Protection Concerns about beneficiaries' privacy and data protection and security associated with private sector involvement in CTP are raised. Financial transfers require that those sending and receiving money must know their customers; however, data privacy and humanitarian regulations require protection of beneficiaries' personal data [15]. In e-transfers, the personal data is more extensive than in in-kind aid and has to be shared with FSPs. Also, the use of the newest technologies that require biometric data, such as iris scanning, is considered as an affront to beneficiaries' human dignity [41]. "We protect our beneficiaries, we do not give names to banks, we give them a number linked to a beneficiary," said a cash expert from WFP. However, it is common practice for HOs to have agreements with the providers on private and confidential data protection of beneficiaries.

The Know Your Customer (KYC) regulations, which are applied to all FSPs, refer to the ID checks that financial institutions perform to comply with national financial regulations. KYC is designed to combat money laundering, terrorist financing, and other related threats to the financial system [42]. HOs are not subject to KYC regulations directly, but via the FSPs. An additional challenge related to KYC regulations is that many beneficiaries often have no identity documents. Thus, HOs have to find alternatives to serve them. FSPs show some understanding to the circumstance that HOs face. However, FSPs are reluctant about risks related to data security. An expert from MasterCard explained, "I think it is a public and private effort to understand how to work together on how to secure data from vulnerable people. From the private sector perspective, I think there is no single company that wants to associate with organizations or situations which are involved in data security issues because it is a reputational issue for us."

Signs of New Donation Flows from Institutional Donors Our interviews also reveal a new trend of fund flows from donors. Intuitional donors, such as ECHO, are willing to contract directly the private sector, instead of via HOs, to execute CTPs. Institutional donors consider that the directly contract with the private sector could enable donated cash to reach beneficiaries faster and inexpensively [43]. As an interviewed cash expert from DRC explained, "what ECHO tries to do is to eliminate financial costs. They are working a lot with UN and that's very inefficient from their perspective. They are spending 3% for the financial service providers and then they add a 13% to 15% to UN to manage these contracts. It's pretty silly. For other HOs, (overhead and plus the operational costs) are a bit less, which is 7%; however, that's still 7% overhead and plus the operational costs in the country. It is understandable why ECHO wants to do it.".

The private sector is positive to this initiative, "we are very excited about this initiative since this is the work that we know how to do it and donors should take advantage of our expertise," highlighted an expert from MasterCard. Issues related to the ownership of the programs and management of the data are related with this direct partnership between institutional donors and FSPs. An interviewed cash expert from DRC said, ". . . there is an issue of the ownership of the data, which is not clear on the guidance of ECHO. UNHCR collects data of refugees, but with the new plan of ECHO, who manages the data? If ECHO contracts directly the private sector, then who is handing the data?" The new trend in the flow of donations from institutional donors directly to FSPs may end up decreasing donations to HOs. A cash expert from NRC said, "we clearly need to communicate to donors and especially to private donors the importance and the value of the other phases of CTP like assessment, targeting and registration to receive funding." This evolution may affect the role of HOs in the humanitarian response.

Conclusions and Future Research

The presented work makes a unique contribution to studies on CTP in that it sheds light on the reciprocal relationship between the private and humanitarian sectors. The private sector's participation is essential for a successful implementation of CTP due to the humanitarian sector's lack of the relevant expertise. Despite the essential role of the private sector in CTP, no study has explored the partnership between these two sectors in CTP. To fill in this research gap, we have first overviewed the mechanism of CTP. Specifically, we have looked into the mechanism from the three key perspectives: stakeholders in CTP, HOs' decision criteria which support their determination on an appropriate CTP option, and CTP actually used in delivering cash to beneficiaries. Furthermore, through a series of interviews, we have discovered challenges related to those three perspectives.

Looking into the observed interaction between the private and humanitarian sectors in CTP, this research unveils challenges, which would hinder partnerships between the two sectors. In addition, our interviews reveal private sector's motivation to be involved with CTP. Regarding the private sector's motivation, interviewed private companies are involved with CTP not only for their pursuance of corporate social responsibility goals, but also for an opportunity to develop robust products and services under various extreme circumstances. Concerning challenges for establishing partnerships in CTP, we identify that the challenges are attributable to five perception gaps (knowledge on finance, governance structure, market competition and services offered, and humanitarian data protection). We have also discussed recent new donation flows accompanied by CTP, in which funds are sent from institutional donors directly to the private sector, rather than via the humanitarian sector.

In closing the paper, we highlight three avenues for future research. We find signs of new donation flows from donors through our interviews. Institutional donors are increasingly willing to contract directly FSPs in CTP because they consider that cash would reach beneficiaries more efficiently. This new trend of donation flows would warrant further research on dynamics between the two sectors as actors is humanitarian response. Secondly, considering that CTP relies on donations collected after a disaster, stakeholders need to consider consequences of liquidity problems. Horizontal cooperation can be observed for in-kind aid by the United Nations Humanitarian Response Depot (UNHRD). The applicability evaluation of a horizontal cooperation system to CTP between HOs is worth exploring. Lastly, we recognize that the hawala system transfers money without a legal contract between the entry and the exit points. What kind of information systems should be designed for the informal cash flow from the perspective of vendors and HOs is a promising future research area.

Acknowledgements We thank our interviewees for providing us with their useful information and insights on their CTP. We also thank anonymous reviewers for their constructive feedbacks on this article. This research was partially funded by the Austrian Science Fund (FWF): Project 26015.

References

1. Ali, D., Gelsdorf, K.: Risk-averse to risk-willing: learning from the 2011 Somalia cash response. Glob. Food Sec. **1**, 57–63 (2012)
2. Harvey, P., Bailey, S.: Good practice review cash transfer programming in emergencies. Humanitarian Practice Network at Overseas Development Institute. Strategic Note Cash Transfers in Humanitarian Contexts. http://odihpn.org/wp-content/uploads/2011/06/gpr11.pdf (2011). Accessed Sept 2017
3. Nurmala, N., De Leeuw, S., Dullaert, W.: Humanitarian–business partnerships in managing humanitarian logistics. Supply Chain Manag. Int. J. **22**(1), 82–94 (2017)
4. Van Wassenhove, L.N.: Blackett memorial lecture humanitarian aid logistics: supply chain management in high gear. J. Oper. Res. Soc. **57**, 475–489 (2006)
5. Adato, M., Bassett, L.: What is the potential of cash transfers to strengthen families affected by HIV and AIDS? A review of the evidence on impacts and key policy debates. Washington DC: Joint Learning Initiative on Children and HIV/AIDS. International Food Policy Research Institute. http://programs.ifpri.org/renewal/pdf/JLICACashTransfers.pdf (2008). Accessed June 2017
6. Doocy, S., Sirois, A., Anderson, J., Tileva, M., Biermann, E., Storey, J.D., Burnham, G.: Food security and humanitarian assistance among displaced Iraqi populations in Jordan and Syria. Soc. Sci. Med. **72**(2), 273–282 (2011)
7. Cash Learning Partnership—CaLP: The state of the world's cash report cash transfer programming in humanitarian aid. http://www.cashlearning.org/downloads/calp-sowc-report-web.pdf (2018). Accessed Feb 2018
8. Cash Learning Partnership—CaLP: Walking the talk: the grand bargain & cash transfer programming. http://www.cashlearning.org/news-and-events/news-and-events/post/459-walking-the-talk-the-grand-bargain%2D%2Dcash-transfer-programming (2017). Accessed May 2017
9. Grand Bargain: "The Grand Bargain: A Shared Commitment to Better Serve People in Need". https://reliefweb.int/sites/reliefweb.int/files/resources/Grand_Bargain_final_22_May_FINAL-2.pdf (2016). Accessed July 2017
10. Heaslip, G., Kovács, G., Haavisto, I.: Cash-based response in relief: the impact for humanitarian logistics. J. Humanit. Logist. Supply Chain Manag. (2018). https://doi.org/10.1108/JHLSCM-08-2017-0043

11. Hoddinott, J., Sandstrom, S., Upton, J.: The impact of cash and food transfers: evidence from a randomized intervention in Niger. IFPRI, Discussion Paper 1341. Washington, DC (2014)
12. Margolies, A., Hoddinott, J.: Costing alternative transfer modalities. J. Dev. Eff. **7**(1), 1–16 (2014)
13. Maunder, N., Dillon, N., Smith, G., Truelove, S., De Bauw, V.: Evaluation of the use of different transfer modalities in ECHO humanitarian aid actions 2011–2014. DG ECHO. http://ec.europa.eu/echo/sites/echosite/files/evaluation_transfer_modalities_final_report_012016_en.pdf (2015). Accessed June 2017
14. Andersson, N., Paredes-Solis, S., Sherr, L., Cockcroft, A.: Cash transfers and social vulnerability in Bosnia: a cross-sectional study of households and listed beneficiaries. Disaster Med. Public Health Prep. **7**, 232–240 (2013)
15. Bailey, S., Gordon, L.: Humanitarian cash transfers and the private sector: background note for the high level panel on humanitarian cash transfer. ODI. https://www.odi.org/publications/9709-humanitarian-cash-transfers-and-private-sector (2015). Accessed July 2017
16. Cozzolino, A., Wankowicz, E., Massaroni, E.: Logistics service providers' engagement in disaster relief initiatives: an exploratory analysis. Int. J. Qual. Serv. Sci. **9**(3/4), 269–291 (2017)
17. Tomasini, R., Van Wassenhove, L.N.: From preparedness to partnerships: case study research on humanitarian logistics. Int. Trans. Oper. Res. **16**, 549–559 (2009)
18. Bealt, J., Barrera, J.C.F., Mansouri, A.: Collaborative relationships between logistics service providers and humanitarian organizations during disaster relief operations. J. Humanit. Logist. Supply Chain Manag. **6**(2), 118–144 (2016)
19. Vega, D., Roussat, C.: Humanitarian logistics: the role of logistics service providers. Int. J. Phys. Distrib. Logist. Manag. **45**(4), 352–375 (2015)
20. Doocy, S., Tappis, H.: Cash-based approaches in humanitarian emergencies: a systematic review. In: 3ie Systematic Review Report 28. International Initiative for Impact Evaluation (3ie), London (2016)
21. Tappis, H., Doocy, S.: The effectiveness and value for money of cash-based humanitarian assistance: a systematic review. J. Dev. Eff. **10**(1), 121–144 (2017)
22. Gelan, A.: Cash or food aid? A general equilibrium analysis for Ethiopia. Dev. Policy Rev. **24**, 601–624 (2006)
23. Simpson, S.M., Parkinson, J., Katsou, E.: Measuring the benefits of using market based approaches to provide water and sanitation in humanitarian contexts. J. Environ. Manag. **216**, 1–7 (2017)
24. De Janvry, A., Finan, F., Sadoulet, E., Vakis, R.: Can conditional cash transfer programs serve as safety nets in keeping children at school and from working when exposed to shocks? J. Dev. Econ. **79**, 349–373 (2006)
25. Filmer, D., Schady, N.: Does more cash in conditional cash transfer programs always lead to larger impacts on school attendance? J. Dev. Econ. **96**(1), 150–157 (2011)
26. Glewwe, P., Kassouf, L.: The impact of the Bolsa Escola/Familia conditional cash transfer program on enrollment, dropout rates and grade promotion in Brazil. J. Dev. Econ. **97**(2), 505–517 (2012)
27. Adato, M., Bassett, L.: Social protection to support vulnerable children and families: the potential of cash transfers to protect education, health and nutrition. AIDS Care Psychol. Sociomed. Aspects AIDS/HIV. **1**, 60–75 (2009)
28. Paxson, C., Schady, N.: Does money matter? The effects of cash transfers on child development in rural ecuador. Econ. Dev. Cult. Change. **59**(1), 187–229 (2010)
29. World Bank: Strategic Note: Cash Transfers in Humanitarian Contexts. International Bank for Reconstruction and Development/The World Bank, Washington (2016)
30. Hoxtell, W., Norz, M., Teicke, K.: Business engagement in humanitarian response and disaster risk management. Global Public Policy Institute, pp. 1–74. http://www.gppi.net/fileadmin/user_upload/media/pub/2015/Hoxtell_et_al_2015_Biz_Engagement_Humanitarian_Repsponse.pdf (2015). Accessed Nov 2017

31. Cash Learning Partnership-CaLP: Glossary of cash transfer programming (CTP) terminology. http://www.cashlearning.org/downloads/calp-updated-glossaryfinal-october-2017.pdf (2017). Accessed Sept 2017
32. Aker, J.A., Boumnijel, R., McClelland, A., Tierney, N: Zap it to me: the short-term impacts of a mobile cash transfer program. CGD Working Paper 268. Washington, D.C. Center for Global Development. http://www.cgdev.org/content/publications/detail/1425470 (2011). Accessed June 2017
33. Creti, P.: Mobile cash transfers for urban refugees in Niamey, Niger. CALP. http://www.cashlearning.org/resources/library/413-mobile-cash-transfers-for-urban-refugees-in-niamey-niger (2014). Accessed Sept 2017
34. Redín, D.M., Calderón, R., Ferrero, I.: Exploring the ethical dimension of Hawala. J. Bus. Ethics. **124**(2), 327–337 (2014)
35. Glewwe, P., Muralidharan, K.: Improving school education outcomes in developing countries: evidence, knowledge gaps, and policy implication. Oxford, UK. http://www.bsg.ox.ac.uk/sites/www.bsg.ox.ac.uk/files/documents/RISE_WP001_Glewwe_Muralidharan.pdf (2015). Accessed Sept 2017
36. UNHCR: Strategy for the institutionalization of cash-based interventions 2016–2020. http://www.unhcr.org/584131cd7.pdf (2016). Accessed Sept 2017
37. De Brauw, A., Gilligan, D., Hoddinott, J., Roy, S.: The impact of bolsa família on women's decision-making power. World Dev. **59**, 487–504 (2014)
38. Soares, V., Silva, E.: Conditional cash transfer programmes and gender vulnerabilities: case studies of Brazil, Chile and Colombia. Overseas Development Institute. https://www.odi.org/sites/odi.org.uk/files/odi-assets/publications-opinion-files/6260.pdf (2010). Accessed Oct 2017
39. Johnson, B., Connolly, E., Carter, S.: Corporate social responsibility: the role of Fortune 100 companies in domestic and international natural disasters. Corp. Soc. Responsib. Environ. Manag. **18**(6), 352–369 (2010)
40. Sandvik, K.B.: Now is the time to deliver: looking for humanitarian innovation's theory of change. J. Int. Humanit. Action. **2**(8), 1–11 (2017)
41. WFP: Guide to personal data protection and privacy. https://docs.wfp.org/api/documents/e8d24e70cc11448383495caca154cb97/download/ (2016). Accessed Nov 2017
42. Elan-Electronic Cash Transfer Learning Action Network-Elan: Data management and protection starter KIT: know you customer regulations. http://elan.cashlearning.org/wp-content/uploads/2016/05/KYC-tipsheet.pdf (2017). Accessed Nov 2017
43. ECHO: Guidance to partners funded by ECHO to deliver medium to large-scale cash transfers in the framework of 2017 HIPs and ESOP. https://ec.europa.eu/echo/sites/echo-site/files/guidance.pdf (2017). Accessed June 2017
44. Andreoni, J.: Philanthrophy. Chapter 18, pp. 1201–1269. In: Kolm, S., Ythier, C., Mercier, E. (eds.). Handbook of the Economics of Giving, Altruism and Reciprocity. http://www.sciencedirect.com/science/article/pii/S1574071406020185 (2006). Accessed Nov 2017
45. Kovács, G.: Where next? The future of humanitarian logistics. In: Christopher, M., Tatham, P. (eds.) Humanitarian Logistics : Meeting the Challenge of Preparing for and Responding to Disasters, 2nd edn, pp. 275–286. Kogan Page, London (2014)
46. MercyCorps: Cash transfer programming-toolkit. https://www.mercycorps.org/sites/default/files/CTP1MethodologyGuide.pdf (2015). Accessed Oct 2017

Equilibrium Analysis for Common-Pool Resources

Lina Mallozzi and Roberta Messalli

Abstract We present an aggregative normal form game to describe the investment decision making situation for a CPR: we will consider a non-cooperative approach searching a Nash equilibrium of it, as well as a cooperative one searching a fully cooperative equilibrium. An application in the Environmental Economics will be illustrated and, in this context, we will introduce a threshold investment as a random variable and we will study the resulting game with aggregative uncertainty looking for a Nash equilibrium and a fully cooperative equilibrium.

Introduction

A common-pool resource (CPR) is a natural or human-made resource, like open-seas fisheries, unfenced grazing range and groundwater basins, from which a group of people can benefit. A CPR consists of two kinds of variables: the stock variable, given by the core resource (for example, in the open-seas fisheries it's the total quantity of fish available in nature), and the flow variable, given by the limited quantity of fringe units that can be extracted (for example, the limited quantity of units that can be fished).

A problem with which a CPR copes is the overuse: in fact, a CPR is a subtractable resource i.e., since its supply is limited, if the quantity that can be restored is used more and more there will be a shortage of it. This problem can lead to the destruction of the CPR (e.g., the CPR is destroyed if, in a short range of period, all the fish of a certain species are taken). Historically, in 1833 the economist William Forster Lloyd published a pamphlet which included an example of overuse of a common parcel of land shared by cattle herders (see [18]).

L. Mallozzi
Department of Mathematics and Applications, University of Naples Federico II, Naples, Italy
e-mail: mallozzi@unina.it

R. Messalli (✉)
Department of Economic and Statistic Science, University of Naples Federico II, Naples, Italy
e-mail: roberta.messalli@unina.it

© Springer Nature Switzerland AG 2018
I. S. Kotsireas et al. (eds.), *Dynamics of Disasters*, Springer Optimization and Its Applications 140, https://doi.org/10.1007/978-3-319-97442-2_4

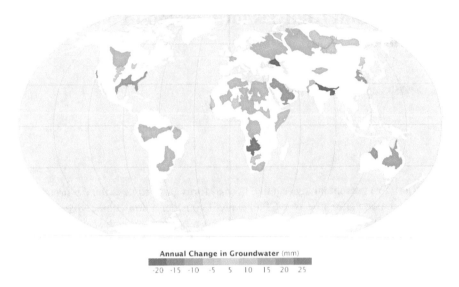

Fig. 1 Acquired January 2003–December 2013. NASA Earth Observatory images by Joshua Stevens using GRACE global groundwater data courtesy of Jay Famiglietti NASA JPL/University of California Irvine and Richey et al. (2015). Caption assembled by Mike Carlowicz, based on stories by Alan Buis (JPL) and Janet Wilson (UCI)

Figure 1 shows the annual change in groundwater storage in the 37 largest groundwater basins in the world: in particular, the basins colored with shades of brown have had more water extracted than the quantity that could be naturally replenished while the ones in blue have had an increase of the level of the water maybe due to precipitation, ice melting, and so on.

In this set, the economic investigation is about how to make a trade-off between preserving and, at the same time, consuming the resource and taking advantage from it.

The concept of CPR goes back to Gordon in 1954 (see [15]) who examines the economic theory of natural resource utilization in the context of the fishing industry.

In order to understand which is the better way to manage a CPR, Hardin in 1968 claims that if all members in a group act just according to their own-self interest with no regards for others, all resources would eventually be destroyed and this leads to the so-called *tragedy of commons* (see [16, 26]), that is a central concept in the study of human ecology and of the environment in general (see [32]).

In 1990, Ostrom lists eight design principles which are prerequisites for a stable CPR management (see [30]). This management is highly dependent on the type of resource considered and in [31] Ostrom points out that the adaptive governance is the best method to obtain a good management of a CPR.

Applications of the CPR concept have been developed later for example in the case of renewable energy (see [35]).

It is important to point out that a CPR is not a Public Good (PG) (see [5]). In fact, they are both non-excludable goods but they differ on the rivalry property in the sense that a PG can be consumed without reducing the availability for others (for example, the air or the national defense), while consuming a CPR will decrease the available resource for others (see [3]).

Aggregative games represent a fundamental instrument to model a game that involves a CPR. This is because a CPR is, as the name itself suggests, a resource that is common for a group of people that benefits from it. Thus, everything that is linked to a CPR depends on the aggregation of strategies and, since the payoff in aggregative games depends explicitly on the aggregation of strategies (see [2]), it is clear that an aggregative game is an appropriate tool in order to model a CPR game and to obtain more sophisticated results.

In this paper we describe an investment decision making situation for a CPR using an aggregative normal form game. For this game we investigate the existence question of a Nash equilibrium solution, that describes situations where agents act in a non-cooperative way. Two types of results are given, with or without convexity-like assumptions. Moreover, in the special case of quadratic return functions, the game is also studied under uncertainty, i.e. when the possibility of a natural disaster with a given probability is considered in the model.

After the introduction, in section Investment in a CPR we describe our model, introduce two different kinds of equilibria, the non-cooperative and the cooperative ones, and show the existence of such equilibria in both cases. Then, we apply these concepts in the context of Environmental Economics in which, for example, players could be countries that choose the level of investment into green policies, in order to be more environmentally friendly (see [33] and the references therein). In section Quadratic Return Under Uncertainty we introduce a threshold investment and we study the resulting game with aggregative uncertainty[1], computing and comparing the non-cooperative and the cooperative equilibria.

Investment in a CPR

In order to describe an investment decision making situation for a CPR let us consider the following normal form game $\langle N, (X_i)_{i=1}^N, (\Pi_i)_{i=1}^N \rangle$ where:

- N is the fixed number of players that can access the CPR;
- for any $i \in \{1, \ldots, N\}$ $X_i = [0, e]$ is the strategy set for player i, where e is the initial endowment that each player can invest in the CPR or in an external activity;

---

[1]Games in which the payoff functions depend on an aggregation of strategy are called *aggregative games* (see [2]) and if, moreover, there is some uncertainty that hits the payoff functions these games are called *aggregative games under uncertainty* (see [20])

- denoted by $X = \prod_{i=1}^{N} X_i$, for any $i \in \{1, \ldots, N\}$ $\Pi_i : X \to \mathbb{R}$ is the payoff function of player i.

Let us denote by $\omega > 0$ the marginal payoff of the external activity and by $x_i \in [0, e]$ the quantity invested by player i in the CPR.

In order to explicitly write the payoff function for each player i, let us introduce a twice continuously differentiable function $G : [0, Ne] \to \mathbb{R}$ that depends on the aggregate invested quantity in the CPR, i.e. $\sum_{i=1}^{N} x_i$, and that is a concave function such that $G(0) = 0$ and $G'(0) > \omega$. This function represents the aggregate return to the investment in the CPR and so the assumption $G'(0) > \omega$ is nothing but an incentive constraint since it means that initially the marginal return to the investment in the CPR is greater than the marginal payoff that everyone achieves if he invests in an external activity (see [14]).

Note that, for any $i \in \{1, \ldots, N\}$, if player i decides to invest part of his endowment in the CPR he will obtain a certain payoff $\omega(e - x_i)$ plus the return of investment in the CPR, $G(\sum_{i=1}^{N} x_i)$, multiplied by the share that is for him, i.e. $x_i / \sum_{i=1}^{N} x_i$, while if he decides to not invest in the CPR he will obtain a payoff ωe. So, let $x = (x_1, \ldots, x_N)$ be a vector of players' investment, then for any $i \in \{1, \ldots, N\}$, the payoff of player i is given by

$$\Pi_i(x_1, \ldots, x_N) = \omega(e - x_i) + \frac{x_i}{\sum_{i=1}^{N} x_i} G\left(\sum_{i=1}^{N} x_i\right). \tag{1}$$

Since $G(0) = 0$ and $G'(0) > 0$, mathematically 0 is a zero of order one for the function G and so $G(\sum_{i=1}^{N} x_i) = \sum_{i=1}^{N} x_i H(\sum_{i=1}^{N} x_i)$ with $H : [0, Ne] \to \mathbb{R}$ such that $H(0) \neq 0$ and H is twice continuously differentiable. So, for any $i \in \{1, \ldots, N\}$, we can rewrite the payoff function of player i, in terms of average aggregate return $H(\sum_{i=1}^{N} x_i)$, in the following way:

$$\Pi_i(x_1, \ldots, x_N) = \omega(e - x_i) + x_i H\left(\sum_{i=1}^{N} x_i\right). \tag{2}$$

Let us explicitly note that, in the context of CPR, the function $G(\sum_{i=1}^{N} x_i)$ is the production function and $H(\sum_{i=1}^{N} x_i)$ is the average production function (see [14]). In the context of Environmental Economics, if the considered strategies are emissions, the function $G(\sum_{i=1}^{N} x_i)$ represents the damage cost function and $H(\sum_{i=1}^{N} x_i)$ its average (see [8, 17]).

Since G is a concave function we can easily obtain, by using Lagrange's theorem, that

$$G'\left(\sum_{i=1}^{N} x_i\right) < \frac{G(\sum_{i=1}^{N} x_i)}{\sum_{i=1}^{N} x_i}, \tag{3}$$

which means that the output elasticity is less than zero.

Using inequality (3), one can check that $H'(\sum_{i=1}^{N} x_i) < 0$ so H is a strictly decreasing function.

In addition, we assume that the following inequality holds:

$$H'\left(\sum_{i=1}^{N} x_i\right) + \sum_{i=1}^{N} x_i H''\left(\sum_{i=1}^{N} x_i\right) < 0. \tag{4}$$

Inequality (4) is satisfied if H is a linear function, if H is a differentiable concave function, or if it is a differentiable convex function with additional assumptions (for example, $H(t) = (t - k)^2$, with $k > 2Ne$).

Remark 1 Inequality (4) has been already used in the context of an oligopolistic market analysis by Okuguchi (see [29]) and by Sherali-Soyster-Murphy (see [34]) for computing Stackelberg-Nash-Cournot equilibria. In both cases the function H is nothing but the inverse demand function $p(Q)$, i.e. the price at which consumers will demand and purchase a quantity Q.

Let us give an interpretation of assumption (4). Let us consider one player, namely player i, that has to face the average production function $H(\cdot)$. This player invests quantity x_i while the other players' aggregate investment is $\sum_{j \neq i} x_j$. Player i's revenue is $x_i H(x_i + \sum_{j \neq i} x_j)$ and so his marginal revenue is $H(x_i + \sum_{j \neq i} x_j) + x_i H'(x_i + \sum_{j \neq i} x_j)$. The rate of change of his marginal revenue with an increase in the other players' aggregate investment is

$$H'\left(x_i + \sum_{j \neq i} x_j\right) + x_i H''\left(x_i + \sum_{j \neq i} x_j\right), \tag{5}$$

by computing the derivative of the marginal revenue $H(x_i + \sum_{j \neq i} x_j) + x_i H'(x_i + \sum_{j \neq i} x_j)$ with respect to $\sum_{j \neq i} x_j$.

Suppose that $\sum_{j \neq i} x_j > 0$ and $H''(x_i + \sum_{j \neq i} x_j) \leq 0$, then

$$H'\left(x_i + \sum_{j \neq i} x_j\right) + x_i H''\left(x_i + \sum_{j \neq i} x_j\right) < 0$$

since $H'(x_i + \sum_{j \neq i} x_j) < 0$.

Suppose that $\sum_{j \neq i} x_j > 0$ and $H''(x_i + \sum_{j \neq i} x_j) > 0$, then

$$H'\left(x_i + \sum_{j \neq i} x_j\right) + x_i H''\left(x_i + \sum_{j \neq i} x_j\right) \leq$$

$$H'\left(x_i + \sum_{j \neq i} x_j\right) + \left(x_i + \sum_{j \neq i} x_j\right) H''\left(x_i + \sum_{j \neq i} x_j\right) < 0$$

since inequality (4) holds.

Finally, if we suppose that $\sum_{j \neq i} x_j = 0$, it's straightforward to obtain the same result.

So (4) implies that, for any level of investment x_i chosen by player i, his marginal revenue is decreasing when the aggregate investment made by all other players is increasing.

Note that, since $G'(0) = H(0) > \omega$ and $H(\cdot)$ is a strictly decreasing function, if the aggregate investment made by all the players significantly increases then the return due to the investment in the CPR becomes negative. Thus the players would be discouraged to move forward with their investments in CPR.

The N players can act non-cooperatively, looking for the so-called *CPRequilibrium*, or cooperatively, looking for the so-called *fullycooperativeCPRequilibrium* [10].

The non-cooperative approach and the cooperative one are taken into account in CPR, Public Goods, Oligopolies, R&D models (see [7, 11, 13, 14]) and, in case of equilibrium uniqueness in both the approaches, the two kinds of equilibria are compared.

Definition 1 A *fullycooperativeCPRequilibrium*, $CPRE^c$, is an N-ple $(x_1^c, \ldots, x_N^c)$ such that, for each player $i \in \{1, \ldots, N\}$, x_i^c solves the following problem:

$$x_i^c \in argmax_{x_i \in X_i} \sum_{j=1}^{N} \Pi_j(x_1^c, \ldots, x_{i-1}^c, x_i, x_{i+1}^c, \ldots, x_N^c).$$

Remark 2 Note that $\sum_{i=1}^{N} \Pi_i(x_1, \ldots, x_N) = \omega e - \omega \sum_{i=1}^{N} x_i + G(\sum_{i=1}^{N} x_i)$ is a continuous function on $X = [0, e]^N$ so, applying Weirstrass theorem, there exists a symmetric fully cooperative CPR equilibrium.

If the agents involved in this CPR management behave in non-cooperative way, they solve a Nash equilibrium problem.

Definition 2 A *CPRequilibrium* (CPRE) is an N-ple $(x_1^*, \ldots, x_N^*)$ such that, for each player $i \in \{1, \ldots, N\}$, x_i^* solves the following problem:

$$x_i^* \in argmax_{x_i \in X_i} \Pi_i(x_1^*, \ldots, x_{i-1}^*, x_i, x_{i+1}^*, \ldots, x_N^*).$$

The following result guarantees the existence of CPR equilibria in a differentiable payoffs framework.

Theorem 1 (Existence) *Let H be a twice continuously differentiable function such that $H(0) \neq 0$ and $H'(\sum_{i=1}^{N} x_i) < 0$. Suppose that inequality (4) holds. Then there exists a symmetric CPRE.*

In order to prove this result, let us first give the following lemma:

Lemma 1 *Let $H(\cdot)$ be the twice differentiable function considered before and let us assume that inequality (4) holds. Then, for each fixed $X_{-i} = \sum_{j \neq i} x_j \geq 0$, the function $K(x_i) = x_i H(x_i + X_{-i})$ is a strictly concave function of x_i over $x_i \geq 0$.*

Proof First of all let us recall that, since G is a concave function, $H(\cdot)$ is a strictly decreasing function.

Let us show that

$$K''(x_i) = 2H'(x_i + X_{-i}) + x_i H''(x_i + X_{-i}) < 0$$

for each $X_{-i} \geq 0$.

Let us suppose $H''(x_i + X_{-i}) \leq 0$, then $K''(x_i) < 0$ since $H'(x_i + X_{-i}) < 0$. Conversely, let us suppose $H''(x_i + X_{-i}) > 0$. So

$$K''(x_i) \leq 2H'(x_i + X_{-i}) + (x_i + X_{-i})H''(x_i + X_{-i}) <$$

$$H'(x_i + X_{-i}) + (x_i + X_{-i})H''(x_i + X_{-i})$$

and, since inequality (4) holds, we have proved the result.

Proof (Theorem 1) Since Lemma 1 holds, then, for any $i \in \{1, \ldots, N\}$, $\Pi_i(x)$ is a strictly concave function in x_i. So $\Pi_i(x)$ is a quasi-concave function in x_i and since for any $i \in \{1, \ldots, N\}$ $\Pi_i(x)$ is continuous and the strategy set X_i is non-empty, closed, and compact, then a symmetric CPR equilibrium exists (see [25]) .

Remark 3 (Nondifferentiable Case) It may happen that the H function is kinked at some level (or levels) of aggregation, due to a different increasing rate of the aggregate return, for example:

$$H(t) = \begin{cases} a - t & \text{if } t \leq \bar{X} \\ a - \frac{(t+\bar{X})}{2} & \text{if } t > \bar{X} \end{cases}$$

that is not differentiable at $t = \bar{X}$.

In this case, in order to obtain an existence result, it is possible to prove that the game has a potential structure (see [19, 24]). Let us recall that a game $\langle N, (X_i)_{i=1}^{N}, (\Pi_i)_{i=1}^{N} \rangle$ is an *ordinal potential game* if there exists a function $P :$ $X \to \mathbb{R}$, called *potential function*, such that for each player $i \in \{1, \ldots, N\}$, each strategy profile $x_{-i} \in \prod_{j \neq i} X_j$ of i's opponents, and each pair $x_i, y_i \in X_i$ of strategies of player i

$$\Pi_i(x_i, x_{-i}) - \Pi_i(y_i, x_{-i}) > 0 \iff P(x_i, x_{-i}) - P(y_i, x_{-i}) > 0.$$

Moreover, in the context of an ordinal potential game, if for all $i \in \{1, \ldots, N\}$ X_i is a compact set and if the potential function P is upper semi-continuous, then a maximum of P over X exists.

In our context, in the case in which the considered function H is continuous, we can define a function $P : X \to \mathbb{R}$ such that

$$P(x_1, \ldots, x_N) := x_1 \ldots x_N (H(\sum_{i=1}^{N} x_i) - \omega).$$

We can easily prove that the considered game Γ is an ordinal potential game. Thus, denoting by $CPR(\Gamma)$ the set of all possible CPR equilibria of Γ, we have that $max_{x \in X} P \subseteq CPR(\Gamma)$ (see [19, 24]) and so, there exists at least one CPR equilibrium. We have then the following theorem.

Theorem 2 (Existence) *If H is a continuous function, then there exists a CPRE.*

Let us explicitly remark that the differentiability assumption is a condition useful in several computational procedures [10], so sometimes the first existence result has to be considered by using differentiable payoff functions.

Quadratic Return Under Uncertainty

In the literature there are several papers dealing with a cooperative as well as a non-cooperative approach to game theoretical models involving Environmental Economics (see, for example, [1, 4, 12, 23] and the references therein). A huge quantity of environmental problems, such as climate change, loss of biodiversity, ozone depletion, the widespread dispersal of persistent pollutants, and many others, involves the commons (for example, forests, energy, industries, water and so on). In numerous situations the considered payoff functions are quadratic functions. Then, in this section we deal with a quadratic return payoff function case.

As done in [14], let us consider the function

$$G(t) = at - bt^2$$

with $G'(0) = a > \omega$.

In this case, $H(t) = a - bt$ and for all $i \in \{1, \ldots, N\}$, the payoff function becomes

$$\Pi_i(x_1, \ldots, x_N) = \omega(e - x_i) + x_i[a - b(x_1 + \cdots + x_N)]$$

that represents the welfare of country i that comprises benefits from investment, deriving from production and consumption of goods, and damages caused by the aggregate investment.

Following a non-cooperative approach, we can easily show that the CPR equilibrium is an N-ple $(x_1^*, \ldots, x_N^*)$ with

$$x_i^* = min\left\{\frac{a - \omega}{b(N + 1)}, e\right\}$$

for each player $i \in \{1, \ldots, N\}$.

Instead, following a cooperative approach, the fully cooperative CPR equilibrium is an N-ple $(x_1^c, \ldots, x_N^c)$ with

$$x_i^c = min\left\{\frac{a - \omega}{2bN}, e\right\}$$

for each player $i \in \{1, \ldots, N\}$.

Comparing the two kinds of equilibrium, we can check in a straightforward way that for each player $i \in \{1, \ldots, N\}$,

$$x_i^c \leq x_i^*$$

and so, cooperating, each player can invest less in the CPR.

In the case of Environmental Economics, sometimes a disaster event may happen that implies a loss in the payoff of any agent. The disaster can have natural causes (earthquakes, floods, etc.) or it may be due to human harm. In both cases investments in the management of resources are very useful. Suppose that a loss is considered in the payoff if the investment is lower than a given upper bound. More precisely, we suppose that there exists a threshold investment, denoted here by $\overline{X}$, that is a random variable since it depends on the probability of a disaster involving the CPR. In line with [4], we suppose that if the aggregate investment is sufficiently large, the payoff functions do not change with respect to the case without uncertainty, otherwise, if the aggregate investment is relative low, every player suffers a loss.

In order to explicitly write the payoff functions, let us fix $\overline{X} \in (0, Ne)$, the critical level, and a constant $L \geq 0$, that represents the loss value. Thus, in the case where $G(t) = at - bt^2$, the payoff functions are

$$\hat{\Pi}_i(x_1, \ldots, x_N) = \begin{cases} \omega(e - x_i) + x_i[a - b(\sum_{i=1}^{N} x_i)] & if \quad \sum_{i=1}^{N} x_i \geq \overline{X} \\ \omega(e - x_i) + x_i[a - b(\sum_{i=1}^{N} x_i)] - L & if \quad \sum_{i=1}^{N} x_i < \overline{X} \end{cases}$$

where player i suffers a loss L when the aggregate invested quantity is below the threshold level.

As in section Investment in a CPR, the N players can act either non-cooperatively, looking for a so-called *CPR equilibrium under uncertainty*, or cooperatively, looking for a so-called *fully cooperative CPR equilibrium under uncertainty*.

The uncertainty is about the threshold $\overline{X}$: in particular, let us assume that the threshold investment is distributed uniformly, i.e. with probability distribution function

$$f(X) = \frac{1}{Ne}$$

with $X = \sum_{i=1}^{N} x_i \in [0, Ne]$ and so the corresponding cumulative distribution function is

$$F(X) = \mathscr{P}(\overline{X} \leq X) = \frac{X}{Ne}$$

with $X \in [0, Ne]$.

If players use a cooperative approach, each of them will maximize the expectation of his own payoff function that is

$$\mathbb{E}(\hat{\Pi}_i) = \omega(e - x_i) + x_i(a - bX) - L(1 - F(X)) =$$

$$= \omega(e - x_i) + x_i(a - bX) - L\left(1 - \frac{X}{Ne}\right)$$

and we can easily show that the CPR equilibrium under uncertainty is an N-ple $(x_1^*, \ldots, x_N^*)$ with

$$x_i^* = min\left\{\frac{a - \omega}{b(N + 1)} + \frac{L}{beN(N + 1)}, e\right\}$$

$\forall i \in \{1, \ldots, N\}$.

If players decide to cooperate, they will maximize the expectation of the joint payoff function that is

$$\mathbb{E}(\hat{\Pi}^c) = N\omega e - \omega X + Xa - bX^2 - LN(1 - F(X)) =$$

$$= N\omega e - \omega X + Xa - bX^2 - LN\left(1 - \frac{X}{Ne}\right).$$

In this other frame, we can easily show that the fully cooperative CPR equilibrium (in this case each payoff is strictly concave in its decision variable) under uncertainty is an N-ple $(x_1^c, \ldots, x_N^c)$ with

$$x_i^c = min\left\{\frac{a - \omega}{2bN} + \frac{L}{2bNe}, e\right\}$$

$\forall i \in \{1, \ldots, N\}$.

We note that the CPR and the fully cooperative CPR equilibria are identical in the case in which $N = 1$. When $N \geq 2$, if we suppose that $L < (a - \omega)e$, we can easily show that

$$x_i^c \leq x_i^*$$

and so, only with a minor disaster, if players decide to cooperate, they can invest less in the CPR.

If agents have additional information on the variability of $\overline{X}$, the threshold investment, then one can also consider different probability distributions, leading to different results.

Conclusions

We have studied an aggregative normal form game to describe an investment decision making situation for a CPR. We have considered two directions in order to solve such kind of a problem: the non-cooperative one, showing an existence result of the corresponding equilibrium, and the cooperative one. In the case of Environmental Economics we have computed and compared the equilibria and then we have introduced a threshold investment and we have studied the resulting game with aggregative uncertainty.

There are many other further possible directions of research. A first direction could be to investigate some other solution concepts for the CPR game, for example in line with the study in [27], one could consider a Nash bargaining problem (see [6, 27, 28]) in which players bargain with each other to decide upon the investment levels in a CPR that they are willing to make. Indeed the bargaining theory can be applied to a variety of real-life situations that comprise also international negotiations linked to environmental and related issues such as carbon emissions trading, biodiversity conservation, and so on.

Another possibility could be using the so-called partial cooperative equilibrium definition (see [9, 21, 22]) since, as it happens in various practical situations, the interaction between agents could be a mixture of non-cooperative and cooperative behavior: one or more coalitions can be formed and players within coalitions cooperate but coalitions act non-cooperatively each other.

All these points will be objects of future research.

Acknowledgements This work has been supported by STAR 2014 (linea 1) "Variational Analysis and Equilibrium Models in Physical and Social Economic Phenomena," University of Naples Federico II, Italy.

We would like to thank the referee for carefully reading our paper and for giving such constructive comments, which helped improving the quality of the paper.

References

1. Abdelaziz, F.B., Brahim, M.B., Zaccour, G.: Strategic investments in R & D and efficiency in the presence of free riders. RAIRO Oper. Res. **50**(3), 611–625 (2016)
2. Acemoglu, D., Jensen, M.K.: Aggregate comparative statics. Games Econ. Behav. **81**, 27–49 (2013)
3. Apesteguia, J., Maier-Rigaud, F.P.: The role of rivalry: public goods versus common pool resources. J. Confl. Resolut. **50**(5), 646–663 (2006)
4. Barrett, S.: Climate treaties and approaching catastrophes. J. Environ. Econ. Manag. **66**(2), 235–250 (2013)
5. Batina, R.G., Ihori, T.: Public Goods: Theories and Evidence. Springer Science and Business Media, Berlin (2005)
6. Binmore, K., Rubinstein, A., Wolinsky, A.: The Nash bargaining solution in economic modelling. RAND J. Econ. 176–188 (1986)
7. Breton, M., Turki, A., Zaccour, G.: Dynamic model of R&D, spillovers, and efficiency of Bertrand and Cournot equilibria. J. Optim. Theory Appl. **123**, 1–25 (2004)

8. Breton, M., Sbragia, L., Zaccour, G.: A dynamic model for international environmental agreements. Environ. Resour. Econ. **45**, 25–48 (2010)
9. Chakrabarti, S., Gilles, R.P., Lazarova, E.A.: Strategic behavior under partial cooperation. Theor. Decis. **71**, 175–193 (2011)
10. Chinchuluun, A., Pardalos, P.M., Migdalas, A., Pitsoulis, L.: Pareto Optimality, Game Theory and Equilibria. Springer Optimization and Its Applications, vol. 17. Springer, New York (2008)
11. D'Aspremont, C., Jacquemin, A.: Cooperative and noncooperative R&D in duopoly with spillovers. Am. Econ. Rev. **78**(5), 1133–1137 (1988)
12. Finus, M.: Game theory and international environmental cooperation: a survey with an application to the kyoto-protocol. Fondazione Eni Enrico Mattei, Milan, Italy (2000)
13. Fischbacher, U., Gachter, S., Fehr, E.: Are people conditionally cooperative? Evidence from a public goods experiment. Econ. Lett. **71**(3), 397–404 (2001)
14. Gardner, R., Walker, J.M.: Probabilistic destruction of common-pool resources: experimental evidence. Econ. J. **102**(414), 1149–1161 (1992)
15. Gordon, H.S.: The economic theory of a common-property resource: the fishery. J. Polit. Econ. **62**(2), 124–142 (1954)
16. Hardin, G.: The tragedy of commons. Science **162**, 1243–1248 (1968)
17. Jorgensen, S., Zaccour, G.: Incentive equilibrium strategies and welfare allocation in a dynamic game of pollution control. Automatica **37**, 29–36 (2001)
18. Lloyd, W.F.: Two Lectures on the Checks to Population. Oxford University, Oxford (1833)
19. Mallozzi, L.: An application of optimization theory to the study of equilibria for games: a survey. Central Eur. J. Oper. Res. 1–17 (2013)
20. Mallozzi, L., Messalli, R.: Multi-leader multi-follower model with aggregative uncertainty. Games **8**(3), 1–14 (2017)
21. Mallozzi, L., Tijs, S.: Conflict and cooperation in symmetric potential games. Int. Game Theory Rev. **10**(3), 245–256 (2008)
22. Mallozzi, L., Tijs, S.: Coordinating choice in partial cooperative equilibrium. Econ. Bull. **29**(2), 1467–1473 (2009)
23. Masoudi, N., Zaccour, G.: Adapting to climate change: is cooperation good for the environment? Econ. Lett. **153**, 1–5 (2017)
24. Monderer, D., Shapley, L.S.: Potential games. Games Econ. Behav. **14**, 124–143 (1996)
25. Moulin, H.: Game Theory for the Social Sciences. NYU Press, New York (1986)
26. Moulin, H. Watts, A.: Two versions of the tragedy of the commons. Rev. Econ. Des. **2**(1), 399–421, (1996)
27. Nagurney, A., Shukla, S.: Multifirm models of cybersecurity investment competition vs. cooperation and network vulnerability. Eur. J. Oper. Res. **260**(2), 588–600 (2017)
28. Nash Jr, J.F.: The bargaining problem. Econometrica **18**(2), 155–162 (1950)
29. Okuguchi, K.: Expectations and Stability in Oligopoly Models. Springer, Heidelberg (1976)
30. Ostrom, E.: Governing the Commons. The Evolution of Institutions for Collective Action. Cambridge University Press, Cambridge (1990)
31. Ostrom, E.: The Challenge of Common Pool Resources. Environ. Sci. Policy Sustain. Dev. **50**(4), 8–21 (2010)
32. Ostrom, E., Dietz, T., Dolsak, N., Stern, P.C., Stonich, S., Weber, E.U.: The Drama of the Commons. Division of Behavioral and Social Sciences and Education. National Academic Press, Washington, DC (2002)
33. Schuller, K., Stankova, K., Thuijsman, F.: Game theory of pollution: national policies and their international effects. Games **8**(3), 1–15 (2017)
34. Sherali, H.D., Soyster, A.L., Murphy, F.H.: Stackelberg-Nash-Cournot equilibria: characterizations and computations. Oper. Res. **31**(2), 253–276 (1983)
35. Wolsink, M.: The research agenda on social acceptance of distributed generation in smart grids: renewable as common pool resources. Renew. Sustain. Energy Rev. **16**(1), 822–835 (2012)

A Multitiered Supply Chain Network Equilibrium Model for Disaster Relief with Capacitated Freight Service Provision

Anna Nagurney

Abstract In this paper, a multitiered supply chain network equilibrium model is constructed, consisting of multiple humanitarian organizations, who seek to purchase services from multiple competing freight (logistic) service providers, for transportation of disaster relief supplies to multiple points of demand for distribution to victims. The freight service providers are faced with capacities associated with the volume of shipments that they can transport. We capture the behavior of the humanitarian organizations who individually minimize the total cost associated with payments for the freight transportation and their transaction costs. We also identify the profit-maximizing behavior of the freight service providers. The governing supply chain network equilibrium conditions are formulated as a variational inequality problem and conditions for existence given. We propose an algorithm for the computation of the equilibrium disaster relief item flows and Lagrange multipliers associated with the freight capacity constraints and provide conditions for convergence. The algorithm is then applied to several numerical examples comprising a case study focusing on an international healthcare crisis. In the case study, we explore the impacts of the addition of a freight service provider as well as that of a humanitarian organization on the profits of freight service providers and on the costs incurred by the humanitarian organizations. The theoretical and numerical results in this paper advance game theory frameworks for humanitarian operations and disaster relief, an area in which there is only a limited literature.

Presented at the 3rd International Conference on Dynamics of Disasters, July 5–9, 2017, Kalamata, Greece.

A. Nagurney (✉)
Department of Operations and Information Management, Isenberg School of Management, University of Massachusetts, Amherst, MA, USA
e-mail: nagurney@isenberg.umass.edu

© Springer Nature Switzerland AG 2018
I. S. Kotsireas et al. (eds.), *Dynamics of Disasters*, Springer Optimization and Its Applications 140, https://doi.org/10.1007/978-3-319-97442-2_5

85

Introduction

Freight service provision is an essential component of disaster relief since only with the effective transportation of the critical needs supplies can the suffering of victims be reduced and lives saved. At the same time, transportation portals and possible routes may be disrupted and severely compromised following a disaster, creating additional challenges for transportation services associated with disaster relief.

Although large humanitarian organizations may have acquired their own freight services and, hence, means of transportation of the needed supplies, which can include, for example, water, food, medicines, shelter items, etc., many humanitarian organizations do not have the financial resources to maintain freight fleets. Hence, they need to purchase such services. Freight service providers, in turn, are profit-maximizers, unlike humanitarian organizations and other nongovernmental organizations (NGOs), which are nonprofits. In addition, they compete among one another to acquire business. Hence, their behavior is distinct from that of humanitarian organizations, who not only must responsibly utilize the financial resources donated to them but also are under pressure to deliver a timely response post disasters. Given the fundamental importance of freight service provision post disasters, costs associated with transportation are second only to personnel for humanitarian organizations (see [30]).

In this paper, we construct what we believe is the first general multitiered supply chain network equilibrium model for disaster relief. The model can handle as many humanitarian organizations as needed by the disaster application under investigation; similarly, the number of freight service providers as well as the number of demand points for distribution of the supplies is not fixed but, rather, is as mandated by the disaster. The cost-minimizing behavior of the individual humanitarian organizations is captured and that of the profit-maximizing freight service providers, who are capacitated. The humanitarian organizations have a fixed amount of supplies that they need delivered to the various points of demand. The governing supply chain network equilibrium conditions are formulated as a variational inequality problem and conditions for existence provided. The solution of the model, for which an algorithm is proposed, yields the equilibrium disaster relief item shipments from the humanitarian organizations via the freight service providers, along with the Lagrange multipliers associated with the freight service providers' capacity constraints. The algorithm decomposes the problem into specially structured network subproblems in the disaster relief item flows, and in the Lagrange multipliers. For the latter, we provide closed form expressions for the iterates. Convergence results are also given. We also demonstrate how to recover the prices the freight service providers charge the humanitarian organizations. In order to demonstrate the efficacy and applicability of the game theory framework, we apply the algorithm to several numerical examples comprising a case study inspired by the international healthcare crisis of the Ebola outbreak in 2014 and 2015.

This paper builds on the work of [16], who constructed a game theory model for disaster relief with a single humanitarian organization and multiple competing freight service providers, also commonly referred to as logistics service providers. The new model in this paper significantly extends the one therein by capturing the behavior of multiple competing humanitarian organizations and by including capacities associated with the freight service providers. Moreover, a distinct algorithm is proposed with less restrictive conditions for convergence, which, nevertheless, exploits, at each iteration, the specially structured underlying network structure associated with the required amounts of disaster relief item shipments between the humanitarian organization nodes and the points of demand. The lineage of supply chain network equilibrium models, which, nevertheless, assumed profit-maximizing decision-makers at each tier of the supply chain network, originated with the paper of [19]. A spectrum of supply chain network equilibrium models, static as well as dynamic, can be found in the book by Nagurney [15]). Supply chain network equilibrium models with a freight sector have recently incorporated price and quality competition among manufacturers and freight service providers (see [22]) and time-based supply chain network competition (see [24]). Here, in contrast, the model developed in this paper includes the humanitarian sector and the quantities demanded are no longer elastic and price-sensitive but, rather, fixed, since, post-disaster, the critically needed product supplies must be delivered.

The paper by Nagurney et al. [17], in turn, considered multiple competing humanitarian organizations engaged in disaster relief, who competed for financial funds and provided needed supplies to the victims and demand points, which were subject to upper and lower bounds. The model was a Generalized Nash Equilibrium model, and, because of the structure of its financial donation functions, was amenable to reformulation as an optimization problem. That model was, subsequently, extended to capture competition on the logistics side and to handle more general financial donation functions using the concept of variational equilibrium by Nagurney et al. [18]. However, in both of these models there was not an explicit tier of freight service providers. Muggy and Heier Stamm [12] provide a thorough review of game theory in humanitarian operations to that date and emphasize that there are many untapped research opportunities for modeling in this area. See also the dissertation of [11]. In these references, however, there are no multitiered supply chain network equilibrium models that include the essential freight service provisioning tier of decision-makers. The relevance of game theory to disaster relief provides new avenues for research, since, principally, centralized decision-making has been modeled using optimization techniques in a variety of settings, especially in the context of transportation, as in evacuation networks (cf. [10, 13, 31, 33, 35, 38], and the references therein), relief routing (cf. [7]), and last mile distribution (see, e.g., [3] and the references therein). Nagurney et al. [20], in turn, developed a supply chain network optimization model for disaster relief under demand uncertainty, whereas [21] also considered cost uncertainty. Both of these models were formulated and solved as variational inequality problems. For additional background on supply chain management and disaster relief, see [37]. Our focus in this paper, in contrast, is on noncooperative game theory (cf. [25, 27]). This framework can also serve in the future as the basis for further research on cooperative game theory as in the case of Nash bargaining solutions (cf. [26, 28]).

For an application of noncooperative game theory and cooperative game theory to cyber security investments, see [23].

This paper is organized as follows. In section The Multitiered Supply Chain Network Equilibrium Model for Disaster Relief, we present the multitiered supply chain network equilibrium model for disaster relief with capacitated, competing freight service providers. The behavior of both the humanitarian organizations and that of the freight service providers is detailed and the governing supply chain network equilibrium conditions defined. The variational inequality formulation is then derived and conditions for existence of an equilibrium solution given. Section The Computational Procedure presents the algorithm and identifies the special network structure of the induced subproblems, along with conditions for convergence. In section An Ebola Case Study the case study is described. The case study is inspired by the recent Ebola healthcare crisis and focuses on the delivery of personal protective equipment (PPEs) needed by the medical professionals who cared for those infected by this highly contagious disease in western Africa. The case study builds on a dataset formulated by Nagurney [16], to which freight service provision capacities are added, as well as another freight service provider and, subsequently, an additional humanitarian organization. Complete results are reported in terms of the equilibrium solutions as well as the total costs incurred by the humanitarian organizations, under the three distinct scenarios, and the profits of the freight service providers. Prices that the humanitarian organizations are charged by the freight service providers are also reported. We discuss the impacts of additional freight service providers and humanitarian organizations as to who wins and who loses. In section Summary and Conclusions, we summarize our results and present our conclusions.

The Multitiered Supply Chain Network Equilibrium Model for Disaster Relief

We consider m humanitarian organizations involved in delivering relief supplies post a disaster, with a typical organization denoted by i. The relief items can be food, water, medicines, shelter supplies, as well as supplies needed by the emergency and healthcare professionals responding to the disaster, etc. There are n competing freight service providers that the organizations can avail themselves of for transporting the relief items, with a typical freight service provider denoted by j. The humanitarian organizations are interested in having the relief items delivered to o points of demand for distribution to the victims, with a typical demand point denoted by k. The multitiered structure of the disaster relief supply chain network is depicted in Figure 1.

The humanitarian organizations compete among themselves for the freight service provision and the freight service providers compete for their business. The humanitarian organizations, which are nonprofits, seek to individually minimize the total costs associated with having their disaster relief supplies delivered by the freight service providers to the victims of the disaster at the demand points. The

Fig. 1 The multitiered
disaster relief humanitarian
organization and freight
service provision supply
chain network

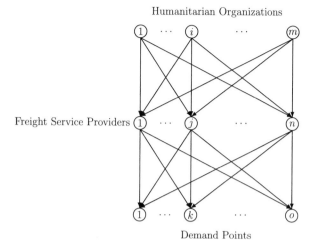

freight service providers, in turn, seek to maximize their profits and compete among one another for the transport of the relief items for the humanitarian organizations.

We first describe the behavior of the humanitarian organizations and then that of the freight service providers. We, subsequently, state the network equilibrium conditions for the disaster relief supply chain network and derive the variational inequality formulation. Qualitative properties of the equilibrium pattern are also given. We assume that all vectors are column vectors.

Behavior of the Humanitarian Organizations

Each humanitarian organization i; $i = 1, \ldots, m$, wishes to have an amount s_k^i of the relief item, which it has in stock and has prepositioned, transported to demand points: $k = 1, \ldots, o$. Let Q_{jk}^i denote the amount of the relief item that i contracts with freight service provider j to have delivered to demand point k. We group the relief item shipments of each humanitarian organization i into the vector $Q^i \in R_+^{no}$.

The per unit price that freight service provider j charges i for transport to k is denoted by ρ_{jk}^{i*}. These prices are revealed once the supply chain network equilibrium model for disaster relief is solved. We demonstrate the procedure of how to recover these prices following the derivation of the variational inequality of the governing supply chain network equilibrium conditions.

Each humanitarian organization i is faced with a total cost $\hat{c}_j^i$ associated with transacting with freight service provider j. This cost includes the cost associated with handling the product until pickup by provider j and interacting with provider j. The total cost $\hat{c}_j^i$; $i = 1, \ldots, m$; $j = 1, \ldots, n$, hence, includes all the costs associated with i contracting with a respective freight service provider j. Observe that the cost associated with a humanitarian organization in transacting with a freight service provider, can, in general, depend not only on its own shipments associated with the freight service provider but also on those of other humanitarian organizations and the same or other freight service providers. The freight service providers guarantee delivery of the disaster relief items in a timely fashion, given what is known about the disaster landscape, and charge accordingly.

The optimization problem faced by humanitarian organization i; $i = 1, \ldots, m$, with the objective function representing total cost to be minimized, is:

$$\text{Minimize} \quad \sum_{j=1}^{n} \sum_{k=1}^{o} \rho_{jk}^{i*} Q_{jk}^{i} + \sum_{j=1}^{n} \hat{c}_{j}^{i}(Q) \tag{1}$$

subject to:

$$\sum_{j=1}^{n} Q_{jk}^{i} = s_{k}^{i}, \quad k = 1, \ldots, o, \tag{2}$$

$$Q_{jk}^{i} \geq 0, \quad j = 1, \ldots, n; k = 1, \ldots, o. \tag{3}$$

The first term preceding the plus sign in the objective function (1) corresponds to the amount that i must pay to the freight service providers whereas the term following the plus sign is the total costs associated with transacting with the freight service providers. Equation (2) guarantees that the relief supplies are delivered to the points of demand. Equation (3) is the nonnegativity assumption for the relief item flows. We define the feasible set K^{i}; $i = 1, \ldots, m$, where $K^{i} \equiv \{Q^{i} | Q^{i} \geq 0$ and satisfies (2)\}. We then define the feasible set $K \equiv \prod_{i=1}^{m} K_{i}$ for all the humanitarian organizations.

Remark In the case that humanitarian organization i; $i = 1, \ldots, m$, has to purchase some or all of the disaster relief supplies then the total cost functions $\hat{c}_{j}^{i}$; $j = 1, \ldots, n$, would include the purchasing cost, in addition to the freight service provision transaction costs.

We assume that the total cost functions $\hat{c}_{j}^{i}$; $i = 1, \ldots, m$; $j = 1, \ldots, n$, are continuously differentiable and convex. Under these assumptions, and the fact that K is convex, we know that a solution to the above optimization problems for the m humanitarian organizations, who compete for freight service provision, simultaneously, coincides with a solution to the variational inequality problem: determine $Q^{*} \in K$, such that

$$\sum_{i=1}^{m} \sum_{j=1}^{n} \sum_{k=1}^{o} \left[\sum_{l=1}^{n} \frac{\partial \hat{c}_{l}^{i}(Q^{*})}{\partial Q_{jk}^{i}} + \rho_{jk}^{i*} \right] \times \left[Q_{jk}^{i} - Q_{jk}^{i*} \right] \geq 0, \quad \forall Q \in K. \tag{4}$$

This result follows from the connection between Nash equilibria (cf. [25, 27]) and variational inequalities (cf. [6] and [14]).

Behavior of the Freight Service Providers

Since the freight service providers are profit-maximizers, they must cover their costs. The cost associated with freight service provider j delivering the relief items from i to demand point k is denoted by c_{ik}^{j}, where, here we assume, for the sake of generality, and in order to effectively capture competition, that

$$c_{ik}^{j} = c_{ik}^{j}(Q), \quad j = 1, \ldots, m, \tag{5}$$

with the freight service provider cost functions assumed to be continuously differentiable and convex. Note that the cost functions in (5) depend, in general, not only on the freight service provider's shipment quantities but also on those of the other freight service providers, since there may be congestion, competition for labor, etc.

In our model each humanitarian organization is providing a similarly sized relief item. Also, each freight service provider can consolidate the shipments from the various humanitarian organizations, if need be, and then transport to points of demand, as inferred by the topology of the network in Figure 1. Each freight service provider j; $j = 1, \ldots, n$, has an associated capacity, denoted by u_j. Hence, the following constraint must hold for each provider j:

$$\sum_{i=1}^{m} \sum_{k=1}^{o} Q_{jk}^{i} \leq u_j. \tag{6}$$

We make the assumption that the total shipment capacity availability is sufficient to meet the total demand, that is,

$$\sum_{i=1}^{m} \sum_{k=1}^{o} s_{k}^{i} \leq \sum_{j=1}^{n} u_j.$$

The optimization problem faced by freight service provider j; $j = 1, \ldots, n$, with the objective function corresponding to the profits to be maximized, is:

$$\text{Maximize} \quad \sum_{i=1}^{m} \sum_{k=1}^{n} \rho_{jk}^{i*} Q_{jk}^{i} - \sum_{i=1}^{m} \sum_{k=1}^{n} c_{ik}^{j}(Q) \tag{7}$$

subject to (6) and:

$$Q_{jk}^{i} \geq 0, \quad k = 1, \ldots, n. \tag{8}$$

As in [16], but in a simpler, single humanitarian organization competitive freight service provider supply chain, and without capacities, we assume that the freight service providers j; $j = 1, \ldots, n$, compete noncooperatively for the disaster relief items, each one seeking to maximize its profits. We associate a nonnegative Lagrange multiplier λ_j with capacity constraint (6) for each freight service provider j; $j = 1, \ldots, n$, and we group the Lagrange multipliers for all freight service providers into the vector $\lambda \in R^n_+$.

The optimality conditions of all freight service providers holding simultaneously, which correspond to a Nash equilibrium, must satisfy the variational inequality problem (cf. [6, 14, 15]): determine $Q^* \in R^{mn}_+$ and $\lambda^* \in R^n_+$, such that:

$$\sum_{j=1}^{n} \left[\sum_{i=1}^{m} \sum_{k=1}^{o} \left[\sum_{h=1}^{m} \sum_{l=1}^{n} \frac{\partial c_{hl}^j(Q^*)}{\partial Q_{jk}^i} - \rho_{jk}^{i*} + \lambda_j^* \right] \right] \times \left[Q_{jk}^i - Q_{jk}^{i*} \right]$$

$$+ \sum_{j=1}^{n} \left[u_j - \sum_{i=1}^{m} \sum_{k=1}^{o} Q_{jk}^{i*} \right] \times \left[\lambda_j - \lambda_j^* \right] \geq 0, \quad \forall Q \in R^{mn}_+, \forall \lambda \in R^n_+. \quad (9)$$

The network equilibrium conditions for the multitiered disaster relief supply chain network model are given below.

Definition 1 (Supply Chain Network Equilibrium for Disaster Relief) A supply chain network equilibrium for disaster relief is said to be established if the disaster relief flows between the two tiers of decision-makers coincide and the flows, prices, and Lagrange multipliers satisfy the sum of variational inequalities (4) and (9).

According to Definition 1, the humanitarian organization and the freight service providers must agree on the amounts of the relief items that they deliver to the demand points. This agreement is accomplished through the prices ρ_{jk}^{i*}; $i = 1, \ldots, m$; $j = 1, \ldots, n$; $k = 1, \ldots, o$. We first present the variational inequality formulation of the supply chain network equilibrium conditions and then discuss how to recover the equilibrium prices.

Theorem 1 (Variational Inequality Formulation of Supply Chain Network Equilibrium for Disaster Relief) *A disaster relief item shipment pattern $Q^* \in K$ and Lagrange multiplier vector $\lambda^* \in R^n_+$ is a supply chain network equilibrium for disaster relief with capacitated freight service provision if and only if it satisfies the variational inequality problem:*

$$\sum_{i=1}^{m} \sum_{j=1}^{n} \sum_{k=1}^{o} \left[\sum_{l=1}^{n} \frac{\partial \hat{c}_l^i(Q^*)}{\partial Q_{jk}^i} + \sum_{h=1}^{m} \sum_{l=1}^{n} \frac{\partial c_{hl}^j(Q^*)}{\partial Q_{jk}^i} + \lambda_j^* \right] \times \left[Q_{jk}^i - Q_{jk}^{i*} \right]$$

$$+ \sum_{j=1}^{n} \left[u_j - \sum_{i=1}^{m} \sum_{k=1}^{o} Q_{jk}^{i*} \right] \times \left[\lambda_j - \lambda_j^* \right] \geq 0, \quad \forall Q \in K, \forall \lambda \in R^n_+. \quad (10)$$

Proof We first establish necessity, that is, if $Q^* \in K$, $\lambda^* \in R^n_+$, is a supply chain network equilibrium according to Definition 1 then it also satisfies variational inequality (10). Indeed, summation of (4) and (9) yields variational inequality (10) with the shipment flows coinciding.

We now establish sufficiency. We rewrite variational inequality (10) as:

$$\sum_{i=1}^{m}\sum_{j=1}^{n}\sum_{k=1}^{o}\left[\sum_{l=1}^{n}\frac{\partial \hat{c}_i^j(Q^*)}{\partial Q_{jk}^i}+\sum_{h=1}^{m}\sum_{l=1}^{n}\frac{\partial c_{hl}^j(Q^*)}{\partial Q_{jk}^i}-\rho_{jk}^{i*}+\rho_{jk}^{i*}+\lambda_j^*\right]\times\left[Q_{jk}^i-Q_{jk}^{i*}\right]$$

$$\sum_{j=1}^{n}\left[u_j-\sum_{i=1}^{m}\sum_{k=1}^{o}Q_{jk}^{i*}\right]\times\left[\lambda_j-\lambda_j^*\right]\geq 0,\quad \forall Q\in K, \lambda\in R_+^n. \qquad (11)$$

But (11) may be expressed as:

$$\sum_{i=1}^{m}\sum_{j=1}^{n}\sum_{k=1}^{o}\left[\sum_{l=1}^{n}\frac{\partial \hat{c}_l^i(Q^*)}{\partial Q_{jk}^i}+\rho_{jk}^{i*}\right]\times\left[Q_{jk}^i-Q_{jk}^{i*}\right]$$

$$+\sum_{j=1}^{n}\sum_{i=1}^{m}\sum_{k=1}^{o}\left[\sum_{h=1}^{m}\sum_{l=1}^{n}\frac{\partial c_{hl}^j(Q^*)}{\partial Q_{jk}^i}-\rho_{jk}^{i*}+\lambda_j^*\right]\times\left[Q_{jk}^i-Q_{jk}^{i*}\right]$$

$$+\sum_{j=1}^{n}\left[u_j-\sum_{i=1}^{m}\sum_{k=1}^{o}Q_{jk}^{i*}\right]\times\left[\lambda_j-\lambda_j^*\right]\geq 0,\quad \forall Q\in K, \forall \lambda\in R_+^n. \qquad (12)$$

(12) corresponds to Definition 1 holding for the prices and shipment pattern $Q^*\in K$ and the vector of Lagrange multipliers $\lambda^*\in R_+^n$. □

Note that in order to recover the equilibrium prices ρ_{jk}^{i*}, $\forall i,j,k$, one sets, according to (9): $\rho_{jk}^{i*}=\sum_{h=1}^{m}\sum_{l=1}^{n}\frac{\partial c_{hl}^j(Q^*)}{\partial Q_{jk}^i}+\lambda_j^*$, $\forall i,j,k$ with $Q_{jk}^{i*}>0$. By setting the freight delivery prices thus, variational inequality (9) holds, so each freight service provider has maximized his profits. Furthermore, we know that the variational inequality (4) governing the humanitarian organizations' noncooperative behavior also holds under these prices.

We now put variational inequality (10) into standard form (cf. [14]): determine $X^*\in\mathcal{K}$, such that

$$\langle F(X^*), X-X^*\rangle\geq 0,\quad \forall X\in\mathcal{K}, \qquad (13)$$

where $F(X)$ is an N-dimensional vector which is a continuous function from $\mathcal{K}$ to R^N, X is an N-dimensional vector, $\mathcal{K}$ is closed and convex, and $\langle\cdot,\cdot\rangle$ denotes the inner product in N-dimensional Euclidean space. We define $\mathcal{K}\equiv K\times R_+^n$, $X\equiv(Q,\lambda)$. Also, we define $F(X)\equiv(F^1(X), F^2(X))$ where $F^1(X)$ consists

of components F_{jk}^i with $F_{jk}^i(X) \equiv \left[\sum_{l=1}^n \frac{\partial \hat{c}_l^j(Q)}{\partial Q_{jk}^i} + \sum_{h=1}^m \sum_{l=1}^n \frac{\partial c_{hl}^j(Q)}{\partial Q_{jk}^i} + \lambda_j \right];$
$i = 1, \ldots, m; \ j = 1, \ldots, n; \ k = 1, \ldots, o.$ Also, $F^2(X)$ consists of components:
$F_j^2(X) \equiv \left[u_j - \sum_{i=1}^m \sum_{k=1}^o Q_{jk}^i \right]; \ j = 1, \ldots, n.$ Here $N = mno + n.$ Then
variational inequality (10) takes on the standard form (13).

We emphasize that, in this paper, we sometimes will express $\langle x, y \rangle$ as $x^T \cdot y,$
where the superscript T denoted transpose.

Qualitative Properties of the Equilibrium Pattern for Disaster Relief

We now turn to the examination of qualitative properties of the equilibrium pattern,
that is, the solution to variational inequality (10), equivalently, (13).

Since the feasible set $\mathcal{K}$ for our model is unbounded, due to the presence of the
Lagrange multipliers, we impose a coercivity condition.

Theorem 2 (Existence of an Equilibrium Pattern) *If the function $F(X)$ in (13)
is coercive, that is,*

$$\lim_{\substack{X \in \mathcal{K} \\ \|X\| \to \infty}} \frac{\langle F(X), X \rangle}{\|X\|} = \infty, \tag{14}$$

then variational inequality (13) has a solution.

Proof Follows from the classical theory of variational inequalities ([8] and [14]).

The Computational Procedure

Before we proceed to our case study, which is on Ebola in western Africa, we discuss
the computational procedure that we will utilize to solve the numerical examples.
The algorithm that we will apply in the next section to compute the solution to
variational inequality (10), using the standard form (13), is the modified projection
method of [9].

The requirements for convergence are that $F(X)$ is monotone and Lipschitz
continuous, since we know that a solution to our model exists. Below we provide
their definitions, for completeness.

Definition 2 (Monotonicity) The function $F(X)$ as in (13) is said to be monotone
on $\mathcal{K}$ if the following property holds:

$$\langle F(X^1) - F(X^2), X^1 - X^2 \rangle \geq 0, \quad \forall X^1, X^2 \in \mathcal{K}. \tag{15}$$

Definition 3 (Lipschitz Continuity) The function $F(X)$ in (13) is said to be Lipschitz continuous on $\mathcal{K}$ if the following property holds:

$$\|F(X^1) - F(X^2)\| \leq L\|X^1 - X^2\|, \quad \forall X^1, X^2 \in \mathcal{K}. \tag{16}$$

Specifically, the statement of the modified projection method is as follows.

The Modified Projection Method

Step 0: Initialization
Initialize with an $X^0 \in \mathcal{K}$. Set $\tau = 1$ and select α, such that $0 < \alpha < \frac{1}{L}$, where L is the Lipschitz constant for the function $F(X)$ in the variational inequality problem (13).

Step 1: Construction and Computation
Compute $\bar{X}^{\tau-1}$ by solving the variational inequality subproblem:

$$\left[\bar{X}^{\tau-1} + (\alpha F(X^{\tau-1}) - X^{\tau-1})\right]^T \cdot \left[X - \bar{X}^{\tau-1}\right] \geq 0, \quad \forall X \in \mathcal{K}. \tag{17}$$

Step 2: Adaptation
Compute X^τ by solving the variational inequality subproblem:

$$\left[X^\tau + (\alpha F(\bar{X}^{\tau-1}) - X^{\tau-1})\right]^T \cdot \left[X - X^\tau\right] \geq 0, \quad \forall X \in \mathcal{K}. \tag{18}$$

Step 3: Convergence Verification
If $|X^\tau - X^{\tau-1}| \leq \epsilon$, for $\epsilon > 0$, a prespecified tolerance, then, stop; else, set $\tau = \tau + 1$ and go to Step 1.

Note that the iterate $\bar{X}^{\tau-1}$ in (17) is actually the solution to the following quadratic programming problem:

$$\text{Minimize} \quad \frac{1}{2}X^T \cdot X + (\alpha F(X^{\tau-1}) - X^{\tau-1})^T \cdot X, \tag{19}$$

subject to: $X \in \mathcal{K}$.

It is straightforward, given the above, to also construct the quadratic programming formulation that will yield the solution $\bar{X}^{\tau-1}$ to variational inequality subproblem (18).

The modified projection method has nice features for our model. In particular, the relief item flows can be computed using an exact equilibration algorithm highlighted in [16], wherein, however, only one humanitarian organization was modeled and there were no capacities associated with the freight service providers. Indeed, the structure of the induced network subproblems for the relief item flows, in both Steps 1 and 2 of the modified projection method, is as depicted in Figure 2. These are

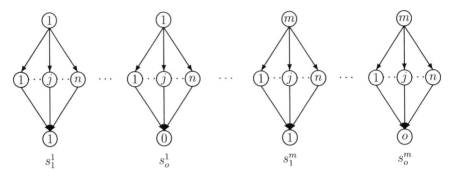

Fig. 2 The special network structure of the relief item subproblems at each iteration of the modified projection method

equivalent to fixed demand transportation network equilibrium problems of special structure (cf. [14]) in that the paths connecting each origin/destination pair of nodes corresponding to the humanitarian organization and demand point pairs have no links in common with any other path.

The Lagrange multipliers, at each iteration, can be solved exactly and in closed form, as detailed below for subproblem (17). An analogous expression can be obtained also for (18).

Explicit Formulae for the Modified Projection Method for the Lagrange Multipliers

The elegance of this algorithm for our variational inequality (10) for the computation of solutions to our model is apparent also from the following explicit formulae, which provide exact solutions for the Lagrange multipliers in subproblem (17). Indeed, we have the following closed form expression for the Lagrange multipliers for $j = 1, \ldots, n$, at iteration $\tau + 1$:

$$\bar{\lambda}_j^{\tau+1} = \max\{0, \lambda_j^{\tau-1} + \alpha(\sum_{i=1}^{m}\sum_{k=1}^{o} Q_{jk}^{i\tau-1} - u_j)\}, \quad j = 1, \ldots, n. \quad (20)$$

For results on the linear convergence rate of the modified projection method, see [35], where references on variants of this algorithm can also be found.

We apply the above modified projection method, with the embedded equilibration algorithm in the next section.

An Ebola Case Study

For our case study, we revisit the Ebola crisis which impacted western Africa in 2014 and 2015. It captured the world's attention because of the suffering of those with the disease and the fear of this highly contagious disease (cf. [2]). This was the worst outbreak of Ebola since it was first identified in 1976. 21 months after the first reported case in March 2014, 11,315 people were reported as having died from Ebola, out of 28,637 cases, in the countries of: Liberia, Sierra Leone, and Guinea, as well as in Nigeria, Mali, and even the US (see [4]). There were eight cases in Nigeria, six in Mali, and one in the US.

In August 2014, according to [1], the World Health Organization declared the Ebola epidemic ravaging parts of west Africa an international health emergency. The [41] reported that over 800 healthcare workers contracted Ebola during this crisis. There were numerous logistical challenges, as well, including that many healthcare facilities had shortages of needed supplies in addition to their workers contracting the disease (see [29]).

Wilson [39] provided a medical professional perspective on this crisis from the frontlines. Wilson [40], in turn, emphasized the importance of logistics and logisticians in battling this disease. Essential items needed by the healthcare workers caring for those stricken with Ebola included personal protective equipment (PPEs), which is the relief item in our case study (see also [5]).

This disease even affected commercial shipping because of the fear of contagion of freight crews (cf. [32]) and, hence, freight provision was under added stress as well as added risk.

Scenario 1: Single Humanitarian Organization, Two Freight Service Providers (Without Capacities and With Capacities), and Three Demand Points

In scenario 1, we consider the supply chain network in Figure 3. There is a single humanitarian organization considering two freight service providers and requiring shipment of the PPEs to each of the three major Ebola impacted countries, that is, to Liberia, Sierra Leone, and Guinea. These countries correspond to demand points 1, 2, and 3, respectively. We first recall the results for the analogous example in [16], which is uncapacitated, but serves as the baseline for our case study, and then we investigate the impacts of the imposition of capacities on the freight service providers.

We utilize the data constructed by Nagurney [16], but here we update our notation to conform to that in this paper. Therein, The World Bank [34] data was used to identify the cost of transport of a container of 20 feet, which can hold 1360 cubic feet of supplies, via ship from the US to these countries. The cost was then multiplied by 14, as per the [36], to obtain an estimated cost for air freight since speed of delivery was essential, given all the existing challenges.

Fig. 3 Supply chain network topology for the Ebola case study scenario 1

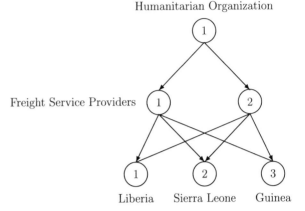

The demands are 10,000 PPE items to each of the three destinations; hence, $s_1^1 = s_2^1 = s_3^1 = 10,000$.

The data are as follows.

The humanitarian organization is faced with the following total costs associated with transacting with the two freight service providers, respectively:

$$\hat{c}_1^1 = 4.50 \times (Q_{11}^1 + Q_{12}^1 + Q_{13}^1), \quad \hat{c}_2^1 = 4.25 \times (Q_{21}^1 + Q_{22}^1 + Q_{23}^1).$$

The humanitarian organization has to purchase the PPE items, so that the $\hat{c}_j^1$; $j = 1, 2$, cost functions include the purchase cost. The total cost associated with freight service provider 1, $\hat{c}_1^1$, is higher than that for freight service provider 2, $\hat{c}_2^1$, since it does not have as much experience with the former provider and the transfer cost is higher per unit.

The freight service provider total costs are as follows:

For freight service provider 1:

$$c_{11}^1 = .0001 Q_{11}^{1~2} + 18.48 Q_{11}^1, \quad c_{12}^1 = .001 Q_{12}^{1~2} + 16.59 Q_{12}^1, \quad c_{13}^1 = .001 Q_{13}^{1~2} + 12.81 Q_{13}^1;$$

For freight service provider 2:

$$c_{11}^2 = .001 Q_{21}^{1~2} + 18.48 Q_{21}^1, \quad c_{12}^2 = .0001 Q_{22}^{1~2} + 16.59 Q_{22}^1. \quad c_{13}^2 = .01 Q_{23}^{1~2} + 12.81 Q_{23}^1.$$

As noted in [16], the nonlinear terms in the cost functions faced by the freight service provider capture the risk associated with transporting the supplies to the points of demand.

The computed equilibrium solution via the projection method, as reported in [16], but adapted here to our new notation, which can handle multiple humanitarian organizations, is:

$$Q_{11}^{1*} = 8,976.31, \quad Q_{12}^{1*} = 796.43, \quad Q_{13}^{1*} = 9,079.99,$$

$$Q_{21}^{1*} = 1,023.69, \quad Q_{22}^{1*} = 9,203.57, \quad Q_{23}^{1*} = 920.01.$$

The prices charged by the freight service providers are:

$$\rho_{11}^{1*} = 20.28, \quad \rho_{12}^{1*} = 18.18, \quad \rho_{13}^{1*} = 30.97,$$

$$\rho_{21}^{1*} = 20.53, \quad \rho_{22}^{1*} = 18.43, \quad \rho_{23}^{1*} = 31.23.$$

The value of the objective function of the humanitarian organization (cf. (1)) is: 829,254.38. The humanitarian organization pays the freight service providers an amount: 697,041.25, which, as noted in [16], corresponds to 84% for transport. This is reasonable since, as also noted in the Introduction, approximately 80% of humanitarian organizations' budgets are towards transportation in disasters. The value of freight service provider 1's objective function (cf. (6)), which coincides with its profits, is: 91,137.94 and that of freight service provider 2 is: 17,982.72. From the results, we see that freight service provider 1 delivers the bulk (the majority) of the PPE supplies to Liberia and Guinea, whereas freight service provider 2 delivers the majority of the supplies to Sierra Leone. Freight service provider 1 carries a total of 18,852.73 of the PPEs whereas freight service provider 2 carries an amount: 11,147.27.

We now assume that upper bounds are imposed on freight service provision with

$$u_1 = 10,000, \quad u_2 = 20,000.$$

In particular, freight service provider 1 has suffered a major disruption in terms of its freight provision in that certain crew members are refusing to deliver the supplies to the Ebola-stricken countries.

Note that the total supply of the PPEs to be delivered is still 30,000 and that is the combined capacity of the two freight service providers. In the original, uncapacitated example, freight service provider 1 delivered almost 19,000 of the PPE items with the remainder being delivered by freight service provider 2.

The modified projection method, as described above, was implemented in FORTRAN and a Linux system at the University of Massachusetts Amherst used for this and the subsequent numerical examples. The algorithm was initialized with s_k^1; $k = 1, 2, 3$, equally divided between the two freight service providers, for each demand point k, to construct the initial disaster relief item shipments. Also, the two Lagrange multipliers associated with the freight service provider capacity constraints were initialized to zero. The convergence tolerance was 10^{-5}, that is, the absolute value of two successive iterates of each of the shipments and each of the Lagrange multipliers differed by no more than this value. We set $\alpha = .3$ in the modified projection method for this scenario.

The modified projection method yielded the following equilibrium shipment and Lagrange multiplier vector solution:

$$Q_{11}^{1*} = 1652.60, \quad Q_{12}^{1*} = 0.00, \quad Q_{13}^{1*} = 8347.40,$$

$$Q_{21}^{1*} = 8347.40, \quad Q_{22}^{1*} = 10000.00, \quad Q_{23}^{1*} = 1652.60,$$

$$\lambda_1^* = 1616.76, \quad \lambda_2^* = 1600.64.$$

The prices charged by the freight service providers are now:

$$\rho_{11}^{1*} = 1635.57, \quad \rho_{12}^{1*} = 1633.35, \quad \rho_{13}^{1*} = 1646.26,$$

$$\rho_{21}^{1*} = 1635.82, \quad \rho_{22}^{1*} = 1619.23, \quad \rho_{23}^{1*} = 1646.50.$$

The humanitarian organization now pays an amount: 49,013,128.00 to the freight service providers. It encumbers a total cost of 49,143,128, which includes its transaction costs. The profit of freight service provider 1 now is: 16,237,542.00 and that of freight service provider 2: 32,119,844.00.

Note that both freight service providers are operating at their respective capacity with freight service provider 1 transporting a total of 10,000 PPEs and freight service provider 2 transporting a total of 20,000 PPEs. Hence, their associated Lagrange multipliers are positive.

Interestingly, the amounts of the PPEs shipped to Liberia have essentially flipped between the two freight service providers as compared to the respective shipment values in the uncapacitated version. Also, interestingly, freight service provider 2 now satisfies the entire demand for PPEs in Sierra Leone, with freight service provider 1 not even servicing this affected country. The prices charged now escalate tremendously because the freight service providers are both at their physical capacities.

Scenario 2: Single Humanitarian Organization, Three Freight Service Providers With Capacities, and Three Demand Points

In scenario 2, the data are as in the capacitated example in scenario 1 except that we add one more freight service provider. Hence, the supply chain network topology is now as in Figure 4. We investigate the impact of enhanced competition among the freight service providers on the humanitarian organization as well as on the original freight service providers.

The added data are as follows.

The cost associated with the humanitarian organization transacting with freight service provider 3 is:

$$\hat{c}_3^1 = 4.75(Q_{31}^1 + Q_{32}^1 + Q_{33}^2)$$

Fig. 4 Supply chain network topology for the Ebola case study scenario 2

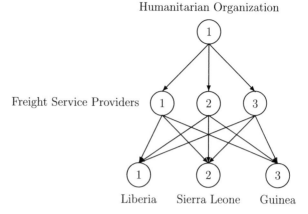

Humanitarian Organization

Freight Service Providers

Liberia   Sierra Leone   Guinea

and the costs associated with freight service provider 3 and the three demand points are:

$$c_{11}^3 = .0001\,Q_{31}^{1\,2} + 12Q_{31}, \quad c_{12}^3 + .0001\,Q_{32}^{1\,2} + 12.5Q_{32}^1, \quad c_{13}^3 = .0001\,Q_{33}^{1\,2} + 11.5Q_{33}^1.$$

According to the above data, the humanitarian organization has higher transaction costs in dealing with the new freight service provider, since it has not done business with it in the past. However, freight service provider 3 is more cost efficient in terms of the three demand points as compared to the original two freight service providers, since it has experience in the western part of Africa.

Also, the capacity of freight service provider 3, $u_3 = 10000$.

We set $\alpha = .1$ in the modified projection method for this example. In order to construct the initial disaster relief item shipments, we divided the supplies needed at each demand point equally among the freight service providers. All three Lagrange multipliers were initialized to zero.

The modified projection method yielded the following equilibrium shipment and Lagrange multiplier vector solution for scenario 2:

$$Q_{11}^{1*} = 5,571.19, \quad Q_{12}^{1*} = 796.68, \quad Q_{13}^{1*} = 3,395.15,$$

$$Q_{21}^{1*} = 682.25, \quad Q_{22}^{1*} = 9,203.32, \quad Q_{23}^{1*} = 351.42,$$

$$Q_{31}^{1*} = 3,746.56, \quad Q_{32}^{1*} = 0.00, \quad Q_{33}^{1*} = 6,253.44.$$

$$\lambda_1^* = 0.00, \quad \lambda_2^* = 0.00, \quad \lambda_3^* = 6.60.$$

The prices charged by the freight service providers are:

$$\rho_{11}^{1*} = 19.59, \quad \rho_{12}^{1*} = 18.18, \quad \rho_{13}^{1*} = 19.60,$$

$$\rho_{21}^{1*} = 19.84, \quad \rho_{22}^{1*} = 18.43, \quad \rho_{23}^{1*} = 19.84,$$

$$\rho_{31}^{1*} = 24.09, \quad \rho_{32}^{1*} = 23.85, \quad \rho_{33}^{1*} = 24.10.$$

Freight service provider 1 transports 9,763.02 PPEs; freight service provider 2, in turn, transports 10,236.98 PPEs, whereas freight service provider 3 transports 10,000.00 PPEs, which is its capacity. Observe that freight service provider 3 charges the highest prices.

The humanitarian organization pays out 621,281.88 to the freight service providers for transportation. It is now faced with a total cost of 756,222.63, which includes its transaction costs. The percentage of total cost for freight is 82%, which is, again, in line with what one sees in practice.

The profit of freight service provider 1 is now: 15,265.55; that of freight service provider 2: 10,170.52, and that of freight service provider 3: 118,765.33. Both original freight service providers suffer financially from enhanced competition. However, the humanitarian organization greatly reduces its total cost. Also, it is interesting to see that freight service provider 3 only transports the PPEs to Liberia and Guinea and delivers no PPE shipments to Sierra Leone. Clearly, humanitarian organizations benefit (and implicitly so do the donors) by having additional freight service providers interested in transporting their relief item cargos.

Scenario 3: Two Humanitarian Organization, Three Freight Service Providers With Capacities, and Three Demand Points

Scenario 3 consists of two humanitarian organization, three freight service providers, and three demand points as illustrated in Figure 5.

In particular, this example has the same data as Scenario 2 but now we add data associated with the second humanitarian organization as detailed below. There is now increased demand for additional PPEs, which the second humanitarian organization is willing to provide.

The second humanitarian organization has worked closely with all the freight service providers in previous disasters and, hence, its transaction costs are lower than those for humanitarian organization 1. The costs associated with the second humanitarian organization transacting with the three freight service providers are:

$$\hat{c}_1^2 = 3(Q_{11}^2 + Q_{12}^2 + Q_{13}^2), \quad \hat{c}_2^2 = 3.5(Q_{21}^2 + Q_{22}^2 + Q_{23}^2), \quad \hat{c}_3^2 = 3(Q_{31}^2 + Q_{32}^2 + Q_{33}^2).$$

The freight service providers, in turn, incur the following costs associated with transporting the disaster relief supplies from humanitarian organization 2:

Fig. 5 Supply chain network topology for the Ebola case study scenario 3

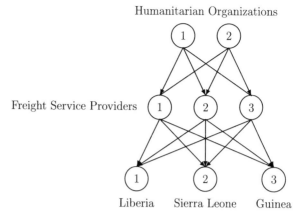

Humanitarian Organizations

Freight Service Providers

Liberia   Sierra Leone   Guinea

$$c_{21}^1 = .0002Q_{11}^2{}^2 + 10Q_{11}^2, \quad c_{22}^1 = .0001Q_{12}^2{}^2 + 8Q_{12}^2, \quad c_{23}^1 = .0002Q_{13}^2{}^2 + 9Q_{13}^2.$$

$$c_{21}^2 = .0001Q_{21}^2{}^2 + 8Q_{21}^2, \quad c_{22}2 = .0002Q_{22}^2{}^2 + 7Q_{22}^2, \quad c_{23}^2 = .0001Q_{23}^2{}^2 + 6Q_{23}^2,$$

$$c_{21}^3 = .0002Q_{31}^2{}^2 + 9Q_{31}^2, \quad c_{22}^3 = .0001Q_{32}^2{}^2 + 7Q_{32}^2, \quad c_{23}^3 = .0001Q_{33}^2{}^2 + 6Q_{33}^2.$$

Also, the amount of the supplies that humanitarian organization 2 wishes to have delivered are:

$$s_1^2 = 3000, \quad s_2^2 = 3000, \quad s_3^2 = 4000.$$

Observe that, now, the total demand for shipments is exactly equal to the total capacity of the three freight service providers.

We, again, set $\alpha = .1$ in the modified projection method. The shipments and Lagrange multipliers were initialized as in Scenario 2 with the former being equally distributed, given the supply/demand, for each humanitarian organization and demand point, among the freight service providers. The modified projection method yielded the following equilibrium shipment and Lagrange multiplier patterns:

$$Q_{11}^{1*} = 5385.70, \quad Q_{12}^{1*} = 689.91, \quad Q_{13}^{1*} = 3415.49,$$

$$Q_{21}^{1*} = 830.67, \quad Q_{22}^{1*} = 9310.09, \quad Q_{23}^{1*} = 368.14,$$

$$Q_{31}^{1*} = 3783.63, \quad Q_{32}^{1*} = 0.00, \quad Q_{33}^{1*} = 6216.37,$$

$$Q_{11}^{2*} = 0.00, \quad Q_{12}^{2*} = 508.91, \quad Q_{13}^{2*} = 0.00,$$

$$Q_{21}^{2*} = 3000.00, \quad Q_{22}^{2*} = 2491.09, \quad Q_{23}^{2*} = 4000.00,$$

$$Q_{31}^{2*} = 0.00, \quad Q_{32}^{2*} = 0.00, \quad Q_{33}^{2*} = 0.00,$$

$$\lambda_1^* = 1569.02, \quad \lambda_2^* = 1568.71, \quad \lambda_3^* = 1575.62.$$

The prices charged by the freight service providers are:

$$\rho_{11}^{1*} = 1588.58, \quad \rho_{12}^{1*} = 1586.99, \quad \rho_{13}^{1*} = 1588.66,$$

$$\rho_{21}^{1*} = 1588.85, \quad \rho_{22}^{1*} = 1587.16, \quad \rho_{23}^{1*} = 1588.88,$$

$$\rho_{31}^{1*} = 1593.13, \quad \rho_{32}^{1*} = 1592.87, \quad \rho_{33}^{1*} = 1593.11,$$

$$\rho_{11}^{2*} = 1579.02, \quad \rho_{12}^{2*} = 1577.12, \quad \rho_{13}^{2*} = 1578.02,$$

$$\rho_{21}^{2*} = 1577.31, \quad \rho_{22}^{2*} = 1576.71, \quad \rho_{23}^{2*} = 1575.51,$$

$$\rho_{31}^{2*} = 1584.62, \quad \rho_{32}^{2*} = 1584.62, \quad \rho_{33}^{2*} = 1581.62.$$

Humanitarian organization 1 pays out 47,689,076.00 to the freight service providers and encumbers a total cost of 47,823,948.00. Humanitarian organization 2 pays out 15,764,292.00 to the freight service providers and encumbers a total cost (which recall includes the transaction costs) of 15,799,038.00. The total disaster relief volume transported by freight service provider 1 is: 10,000.00; the total amount transported by freight service provider 2 is: 20,000.00, and the total amount by freight service provider 3 is: 10,000.00. Hence, all freight service providers are at their respective capacity and, therefore, the Lagrange multipliers are all positive. Freight service provider 1 now enjoys a profit of 15,705,270.00, whereas freight service provider 2 has a profit of 31,388,628.00, and freight service provider 3 a profit of 15,809,006.00.

It is interesting that humanitarian organization 2 does not utilize the services of freight service provider 3 at all and that the majority of its shipments go via freight service provider 2. Humanitarian organization 1, on the other hand, relies primarily on the services of freight service provider 1 for shipments to Liberia and Guinea and the services of freight service provider 2 for shipments to Sierra Leone. With increased demand for their services, because of the needs of humanitarian organization 2, all freight service providers have higher profits than in Scenario 2. Because of the increased competition for freight service provision from the added humanitarian organization, humanitarian organization 1 now has a substantially higher total cost than in Scenario 2.

This example vividly illustrates that the humanitarian organizations might benefit from cooperating rather than competing.

Summary and Conclusions

In this paper, a multitiered supply chain network equilibrium model for disaster relief was constructed, which can handle as many humanitarian organizations as well as freight (logistic) service providers engaged in the delivery of disaster relief supplies to multiple demand points for distribution to the victims, as needed by the specific disaster relief application. In addition, the model incorporates capacities associated with the freight service providers' transportation of the relief items. Previous freight service competitive modeling in disaster relief considered only a single humanitarian organization and was uncapacitated (cf. [16]). Hence, the new model is a significant extension on prior work and also a contribution to the still very limited literature on game theory and disaster relief. Also, although supply chain network equilibrium models have generated a rich literature, beginning with the first model of [19], all, except for the above-noted work and this paper, have focused on profit-maximization as the primary objective of the various decision-makers associated with the supply chain network tiers. Here, in contrast, since we are dealing with humanitarian organizations, which are nonprofit entities, their objective functions are comprised of cost minimization. Of course, the cost can also represent a generalized cost, since these functions are nonlinear, and can capture risk, for example, or even time, weighted accordingly.

In this paper, in addition to the new model, existence results were provided, as well as an algorithm, which has less restrictive conditions for convergence than the previously proposed projection method for the single humanitarian organization case and also results in a decomposition of the disaster relief item flows at each step into network subproblems of special structure and closed form expressions for the Lagrange multipliers associated with the capacity constraints. The algorithm was then applied to a case study inspired by a major international healthcare crisis—that of the Ebola outbreak, which devastated multiple western African countries in 2014 and 2015. In particular, we first included capacities on freight service provision from a dataset constructed in [16] associated with the transportation of personal protective equipment needed by the medical professionals battling Ebola. We then investigated the impact of the addition of a new freight service provider and, subsequently, also the addition of a second humanitarian organization. We reported the equilibrium disaster relief shipments for the three scenarios along with the Lagrange multipliers as well as the payouts of the humanitarian organizations to the freight service providers and their total costs, plus the profits of the freight service providers and the prices that they charge the humanitarian organizations for freight service provision.

The numerical examples making up the case study demonstrate that humanitarian organizations benefit from the availability of a larger number of competitive freight service providers (although this affects freight service providers negatively in terms of profits). Also, the addition of humanitarian organizations competing for services from the freight service providers results in higher prices since the capacities may be achieved. Hence, the case study illustrates that cooperation may be a fruitful avenue for future research on game theory and freight service provision for humanitarian organizations in disaster relief. In addition, it would be interesting to compute solutions to large-scale numerical examples and to explore alternative algorithms for computational purposes. We leave such research for the future.

Acknowledgements The author acknowledges support by the Radcliffe Institute for Advanced Study at Harvard University during the summer of 2017, where she was a Summer Fellow. She also acknowledges support from the John F. Smith Memorial Fund at the University of Massachusetts Amherst.

The author thanks the co-organizers of the 3rd Dynamics of Disasters conference and the participants for engaging discussions. She also thanks the anonymous reviewer for helpful comments and suggestions on an earlier version of this paper.

References

1. Agence France-Presse: UN health agency declares killer Ebola epidenic a global disaster. Daily Nation, 8 Aug 2014. Available at: http://www.nation.co.ke/news/africa/UN-health-agency-declares-killer-Ebola-global-disaster/1066-2413066-11mwufwz/index.html
2. Ap, T.: Ebola crisis: WHO slammed by Harvard-convened over slow response. CNN, 23 Nov 2015
3. Balcik, B., Beamon, B.M., Smilowitz, K.: Last mile distribution in humanitarian relief. J. Intell. Transp. Syst. **12**(2), 51–63 (2008)
4. BBC.com: Ebola: Mapping the outbreak, 14 Jan 2016. Available at: http://www.bbc.com/news/world-africa-28755033
5. Fischer II, W.A., Hynes, N.A., Perl, T.M.: Protecting healthcare workers from Ebola: personal protective equipment is critical but not enough. Ann. Intern. Med. **161**(10), 753–754 (2014)
6. Gabay, D., Moulin, H.: On the uniqueness and stability of Nash equilibria in noncooperative games. In: Bensoussan, A., Kleindorfer, P., Tapiero, C.S. (eds.) Applied Stochastic Control of Econometrics and Management Science, pp. 271–294. North-Holland, Amsterdam (1980)
7. Huang, M., Smilowitz, K., Balcik, B. Models for relief routing: equity, efficiency and efficacy. Transp. Res. E **48**, 2–18 (2012)
8. Kinderlehrer, D., Stampacchia, G.: An Introduction to Variational Inequalities and Their Applications. Academic Press, New York (1980)
9. Korpelevich G.M.: The extragradient method for finding saddle points and other problems. Matekon **13**, 35–49 (1977)
10. Miller-Hooks, E., Sorrel, G.: The maximal dynamic expected flows problem for emergency evacuation planning. Transp. Res. Rec. **2089**, 26–34 (2008)
11. Muggy, L.: Quantifying and Mitigating Decentralized Decision Making in Humanitarian Logistics Systems. PhD Dissertation, Kansas State University, Manhattan (2015)
12. Muggy L., Heier Stamm, J.L.: Game theory applications in humanitarian operations: a review. J. Humanitarian Logistics Supply Chain Manag. **4**(1), 4–23 (2014)
13. Na, H.S., Banerjee, A.: A disaster evacuation network model for transporting multiple priority evacuees. IIE Trans. **47**(11), 1287–1299 (2015)
14. Nagurney, A.: Network Economics: A Variational Inequality Approach, 2nd and rev. ed. Kluwer Academic Publishers, Dordrecht (1999)
15. Nagurney, A.: Supply Chain Network Economics: Dynamics of Prices, Flows, and Profits. Edward Elgar Publishing. Cheltenham (2006)
16. Nagurney, A.: Freight service provision for disaster relief: a competitive network model with computations. In: Kotsireas, I.S., Nagurney, A., Pardalos, P.M. (eds.) Dynamics of Disasters: Key Concepts, Models, Algorithms, and Insights, pp 207–229. Springer International Publishing, Switzerland (2016)

17. Nagurney, A., Alvarez Flores, E., Soylu, C.: A generalized Nash equilibrium model for post-disaster humanitarian relief. Transp. Res. E **95**, 1–18 (2016)
18. Nagurney, A., Daniele, P., Alvarez Flores, E., Caruso, A.: A variational equilibrium framework for humanitarian organizations in disaster relief: effective product delivery under competition for financial funds. In: Kotsireas, I.S., Nagurney, A., Pardalos, P.M. (eds.) Dynamics of Disasters: Algorithmic Approaches and Applications. Springer International Publishers, Switzerland (2018)
19. Nagurney, A., Dong, J., Zhang, D.: A supply chain network equilibrium model. Transp. Res. E **38**, 281–303 (2002)
20. Nagurney, A., Masoumi, A.H., Yu, M.: An integrated disaster relief supply chain network model with time targets and demand uncertainty. In: Nijkamp, P., Rose, A., Kourtit, K. (eds.) Regional Science Matters: Studies Dedicated to Walter Isard, pp. 287–318. Springer International Publishing, Switzerland (2015)
21. Nagurney, A., Nagurney, L.S.: A mean-variance disaster relief supply chain network model for risk reduction with stochastic link costs, time targets, and demand uncertainty. In: Kotsireas, I.S., Nagurney, A., Pardalos, P.M. (eds.) Dynamics of Disasters: Key Concepts, Models, Algorithms, and Insights, pp. 75–99. Springer International Publishing, Switzerland (2016)
22. Nagurney, A., Saberi, S., Shivani, S., Floden, J.: Supply chain network competition in price and quality with multiple manufacturers and freight service providers. Transp. Res. E **77**, 248–267 (2015)
23. Nagurney, A., Shukla, S.: Multifirm models of cybersecurity investment competition vs. cooperation and network vulnerability. Eur. J. Oper. Res. **260**(2), 588–600 (2017)
24. Nagurney, A., Yu, M., Floden, J., Nagurney, L.S.: Supply chain network competition in time-sensitive markets. Transp. Res. E **70**, 112–127 (2014)
25. Nash J.F.: Equilibrium points in n-person games. Proc. Natl. Acad. Sci. U. S. A. **36**, 48–49 (1950)
26. Nash J.F.: The bargaining problem. Econometrica **18**, 155–162 (1950)
27. Nash J.F.: Noncooperative games. Ann. Math. **54**, 286–298 (1951)
28. Nash J.F.: Two person cooperative games. Econometrica **21**, 128–140 (1953)
29. O'Byrne, R.: The Ebola outbreak and why logistics are struggling to cope. Logistics Bureau, 3 Nov 2014. Available at: http://www.logisticsbureau.com/the-ebola-outbreak-and-why-logistics-are-struggling/
30. Pedraza Martinez, A.J., Stapleton, O., Van Wassenhove, L.N.: Field vehicle fleet management in humanitarian operations: a case-based approach. J. Oper. Manag. **29**(5), 404–421 (2011)
31. Regnier, E.: Public evacuation decisions and hurricane track uncertainty. Manag. Sci. **54**(2), 16–28 (2008)
32. Saul, J.: Shipping industry modified clauses in contracts for Ebola risks. www.reuters.com, 16 Dec 2014. Available at: http://www.reuters.com/article/health-ebola-shipping-idUSL6N0U024W20141216
33. Sheffi, Y., Mahmassani, H., Powell, W.B.: A transportation network evacuation model. Transp. Res. A **16**(3), 209–218 (1982)
34. The World Bank: Cost to export (US$ per container) (2016). http://data.worldbank.org/indicator/IC.EXP.COST.CD
35. Tseng, P.: On linear convergence of iterative methods for the variational inequality problem. J. Comput. Appl. Math. **60**, 237–252 (1995)
36. United States Department of Commerce: Access costs everywhere (2016). http://acetool.commerce.gov/shipping
37. Van Wassenhove, L.N. Blackett memorial lecture. Humanitarian aid logistics: supply chain management in high gear. J. Oper. Res. Soc. **57**(5), 475–489 (2006)
38. Vogiatzis, C., Pardalos, P.M.: Evacuation modeling and betweenness centrality. In: Kotsireas, I.S., Nagurney, A., Pardalos, P.M. (eds.) Dynamics of Disasters: Key Concepts, Models, Algorithms, and Insights, pp. 345–350. Springer International Publishing, Switzerland (2016)
39. Wilson, D.: CE: Inside an Ebola ET: A nurses' report. Am. J. Nurs. **115**(12), 28–38 (2015)

40. Wilson, D.: Ode to the humanitarian logistician: humanistic logistics through a nurse's eye. In: Kotsireas, I.S., Nagurney, A., Pardalos, P.M. (eds.) Dynamics of Disasters: Key Concepts, Models, Algorithms, and Insights, pp. 361–369. Springer International Publishing, Switzerland (2016)
41. World Health Organization: Health worker Ebola infections in Guinea, Liberia and Sierra Leone: A preliminary report. Geneva, Switzerland, 21 May (2015)

A Variational Equilibrium Network Framework for Humanitarian Organizations in Disaster Relief: Effective Product Delivery Under Competition for Financial Funds

Anna Nagurney, Patrizia Daniele, Emilio Alvarez Flores, and Valeria Caruso

Abstract In this paper, we present a new Generalized Nash Equilibrium (GNE) model for post-disaster humanitarian relief by introducing novel financial funding functions and altruism functions, and by also capturing competition on the logistics side among humanitarian organizations. The common, that is, the shared, constraints associated with the relief item deliveries at points of need are imposed by an upper level humanitarian organization or regulatory body and consist of lower and upper bounds to ensure the effective delivery of the estimated volumes of supplies to the victims of the disaster. We identify the network structure of the problem, with logistical and financial flows, and propose a variational equilibrium framework, which allows us to then formulate, analyze, and solve the model using the theory of variational inequalities (rather than quasivariational inequality theory). We then utilize Lagrange analysis and investigate qualitatively the humanitarian organizations' marginal utilities if and when the equilibrium relief item flows are (or are not) at the imposed demand point bounds. We illustrate the game theory model through a case study focused on tornadoes hitting western Massachusetts, a highly unusual event that occurred in 2011. This work significantly extends the original model of Nagurney (Dynamics of Disasters: Key Concepts, Models, Algorithms,

Presented at the 3rd International Conference on Dynamics of Disasters Kalamata, Greece, July 5–9, 2017.

A. Nagurney (✉)
Department of Operations and Information Management, Isenberg School of Management, University of Massachusetts, Amherst, MA, USA
e-mail: nagurney@isenberg.umass.edu

P. Daniele · V. Caruso
Department of Mathematics and Computer Science, University of Catania, Catania, Italy

E. A. Flores
Cisco Services, Research Triangle Park, NC, USA

I. S. Kotsireas et al. (eds.), *Dynamics of Disasters*, Springer Optimization and Its Applications 140, https://doi.org/10.1007/978-3-319-97442-2_6

and Insights. Springer International Publishing, Switzerland, 2016), which, under the imposed assumptions therein, allowed for an optimization formulation, and adds to the literature of game theory and disaster relief, which is nascent.

Introduction

Disaster relief is fraught with many challenges: the infrastructure, from transportation to communications to energy delivery, may be damaged or destroyed, and services, from healthcare to governmental ones, impacted, all while victims are in desperate need of relief items such as water, food, medicines, and shelter. A timely response to a disaster, hence, can save lives, reduce suffering, and assist in recovery. Moreover, it can also enhance the reputations of humanitarian organizations and their very sustainability in terms of financial donations.

Disasters come in many forms, from natural disasters, such as tornadoes, earthquakes, and typhoons, which are often sudden-onset, to famines, which are slow-onset, and can occur not only from changes in weather patterns, resulting in droughts, for example, but also from political situations, including war (cf. [44]). Hence, certain disasters are man-made, as in the case of the Syrian refugee crisis (cf. [41]), and terrorist attacks, such as 9/11 ([7]).

The number of disasters is growing as well as the number of people affected by them ([31]) with additional pressures coming from climate change, increasing growth of populations in urban environments, and the spread of diseases brought about by global air travel. The associated costs of the damage and losses due to natural disasters is estimated at an average $117 billion a year between 1991 and 2010 ([46]). Notable sudden-onset natural disasters have included Hurricane Katrina in 2005, which was the second costliest natural disaster in the US, with Hurricane Harvey, which hit Texas in 2017 being the costliest; the Haiti earthquake in 2010, the triple disaster in Fukushima, Japan in 2011, consisting of an earthquake, followed by a tsunami and a nuclear meltdown technological disaster, Superstorm Sandy in 2012, tropical cyclone Haiyan in 2013, which was the strongest cyclone ever recorded, the earthquake in Nepal in 2015, and Hurricane Matthew in 2016. In 2017, the United States experienced a record year in weather and climate disasters, with the cumulative cost exceeding $300 billion ([38]) and with 16 natural disasters costing more than $1 billion each in terms of damages.

The challenges to disaster relief (humanitarian) organizations, including non-governmental organizations (NGOs), are immense. The majority operate under a single, common, humanitarian principle of protecting the vulnerable, reducing suffering, and supporting the quality of life, while, at the same time, competing for financial funds from donors to ensure their own sustainability. As noted in [28], competition is intense, with the number of registered US nonprofit organizations increasing from 12,000 in 1940 to more than 1.5 million in 2012. Approximately $300 billion are donated to charities in the United States each year ([49]). At the same time, many stakeholders believe that humanitarian aid has not been as successful in delivering on the humanitarian principle as might be feasible due to a lack of coordination, which results in duplication of services (see [20]).

We believe that some of the challenges that humanitarian organizations engaged in disaster relief are faced with can be addressed through the use of game theory. Game theory is a methodological framework that captures complex interactions among competing decision-makers (noncooperative games) or cooperating ones (cooperative games). The contributions of John [35, 36], in particular, are highly relevant and established some of the foundations of game theory. Specifically, we note that, in the case of noncooperative games, in which the utilities of the competing players, that is, the decision-makers, in the game, depend on the other players' strategies, the governing concept is that of Nash Equilibrium. If, however, the feasible sets, that is, the constraints, are not specific to each player, but, rather, depend also on the strategies of the other players, then we are dealing with a Generalized Nash Equilibrium, introduced by Debreu [10] (see, also, [13, 45], and the references therein).

In particular, in this paper, we construct a new Generalized Nash Equilibrium (GNE) network model for disaster relief, which models competition among NGOs for financial funds post-disaster, as well as for the delivery of relief items. The utility function that each NGO seeks to maximize depends on its financial gain from donations plus the weighted benefit accrued from doing good through the delivery of relief items minus the total cost associated with the logistics of delivering the relief items. The model extends the earlier model of [33] in the following significant ways, which means that the optimization reformulation, as done in that paper, no longer applies:

1. The financial funds functions, which capture the amount of donations to each NGO, given their visibility through media of the supplies of relief items delivered at demand points, and under competition, need not take on a particular structure.
2. The altruism or benefit functions, also included in each NGO's utility function, need not be linear.
3. The competition associated with logistics is captured through total cost functions that depend not only on a particular NGO's relief item shipments but also on those of the other NGOs.

In order to guarantee effective product delivery at the demand points, we retain the lower and upper bounds, as introduced in [33]. Such common, or shared, constraints, assist in coordination (cf. [1]) and would be imposed by a higher level humanitarian organization or regulatory authority in order to ensure that the needed volumes of relief items are delivered but are not oversupplied, which can result in congestion, materiel convergence, and wastage. We assume that the NGOs have prepositioned the supplies of the disaster relief items and that the total amount available across all NGOs is sufficient to meet the needs of the victims.

It is important to emphasize that Generalized Nash Equilibrium problems are more challenging to formulate and solve and are usually tackled via quasivariational inequalities (cf. [3]), the theory of which, as well as the associated computational procedures, are not in as an advanced state as that of variational inequalities (see [19] and [26]). Here we utilize, for the first time, in the context of humanitarian operations and disaster relief, a *variational equilibrium*. As noted in [33]), a variational equilibrium is a specific kind of GNE (cf. [12, 21]). The variational equilibrium allows for alternative variational inequality formulations of our new Generalized Nash Equilibrium network model. What is notable about a variational equilibrium (see also [22]) is that the Lagrange multipliers associated with the shared or coupling constraints of the NGOs are the same for all NGOs in the disaster relief game. This also provides us with an elegant economic and equity interpretation.

The only other game theory model for disaster relief that includes elements of logistics plus financial funds is that of [28]. Zhuang et al. [49] proposed a model that showed that the amount of charitable contributions made by donors is positively dependent on the amount of disclosure by the NGOs. The authors emphasized that there is a dearth of existing game-theoretic research on nonprofit organizations. Toyasaki and Wakolbinger [42] developed game theory models to analyze whether an NGO should establish a special fund after a disaster (in terms of earmarked donations) or allow only unearmarked donations. Nagurney [28], in turn, presented a network game theory model in which multiple freight service providers are engaged in competition to acquire the business of carrying disaster relief supplies of a humanitarian organization in the amounts desired to the destinations. Coles and Zhuang [6], on the other hand, argued for the need for cooperative game theory models for disaster recovery operations by highlighting a stream of post-disaster operations. Muggy and Heier Stamm [25] give an excellent review of game theory in humanitarian operations and note that there are many untapped research opportunities for modeling in this area. See also the dissertation of [24]. The research in our paper adds to the still nascent literature on game theory and disaster relief / humanitarian operations.

This paper is organized as follows. In Section The Variational Equilibrium Network Framework for Humanitarian Organizations in Disaster Relief, we construct the novel Generalized Nash Equilibrium model for disaster relief, which captures competition both on the financial funds side and on the logistics side and we identify the network structure. We present the variational equilibrium framework and also prove the existence of an equilibrium solution. In addition, we provide, for completeness, the variational inequality formulation of a special case of the model, under the Nash equilibrium solution, in the absence of imposed common demand constraints. In Section Lagrange Theory and Analysis of the Marginal Utilities, we then explore, through Lagrange analysis, the humanitarian organizations' marginal utilities when the equilibrium disaster relief flows are at the upper or the lower bounds of the imposed demands of the regulatory body or lie in between. In order to illustrate the framework developed here, Section The Algorithm and Case Study presents both an algorithmic scheme and a case study, inspired by tornadoes that hit western Massachusetts in June 2011, with devastating impact (cf. [47]). We summarize our results and present our conclusions in Section Summary and Conclusions. We also provide directions for future research.

The Variational Equilibrium Network Framework for Humanitarian Organizations in Disaster Relief

We now present the new Generalized Nash Equilibrium model for disaster relief, along with the variational equilibrium framework. As mentioned in the Introduction, the model extends the earlier model of [33], which, under the imposed assumptions therein, allowed for an optimization reformulation. Our notation follows closely the notation in the above paper but here we utilize, in contrast, a more general variational equilibrium framework.

We consider m humanitarian organizations, here referred to as nongovernmental organizations (NGOs), with a typical NGO denoted by i, seeking to deliver relief supplies, post a disaster, to n demand points, with a typical demand point denoted by j. The relief supplies can be water, food, or medicine. We assume that the product delivered can be viewed as being homogeneous. We denote the volume of the relief item shipment (flow) delivered by NGO i to demand point j by q_{ij}. We group the nonnegative relief item flows from each NGO i; $i = 1, \ldots, m$, into the vector $q_i \in R_+^n$ and then we group the relief item flows of all the NGOs to all the demand points into the vector $q \in R_+^{mn}$. The vector q_i is the vector of strategies of NGO i.

The NGOs compete for financial funds from donors, while also engaging in competition on the logistic side in terms of costs, since there may be competition for freight services, etc., as well as congestion at the demand sites. The network structure of the problem is given in Figure 1. Note that the links from the first tier nodes representing the NGOs to the bottom tier nodes, corresponding to the demand points, are the shipment links and have relief item flows associated with them. The links from the demand nodes to the NGO nodes (in the opposite direction) are the links with the financial flows from the donors reacting to the visibility of the NGOs in their delivery of the needed supplies through the media. The network structure of this problem differs from the network underlying the model given in [28] since in that model, the financial flows, once collected, were partitioned to each NGO, using a factor representing the portion of the financial funds each humanitarian organization was (likely) to get of the total amount donated.

We emphasize that, in terms of the sequence of events, the humanitarian organizations first decide on the level of relief items to be provided at each demand point and deliver the amounts. Then they receive the corresponding financial flows. Therefore, the financial flows are received after the supplies arrive. As noted in [28], empirically, these funds are realized and made available quickly, and these two events are almost concurrent in many cases. The justification of this assumption is also provided by the nature of the incentives of the decision-makers in our model, which is to provide humanitarian relief as quickly as possible whenever a disaster strikes.

Each NGO i encumbers a cost, c_{ij}, associated with shipping the relief items to location j, where we assume that

$$c_{ij} = c_{ij}(q), \quad j = 1, \ldots n, \tag{1}$$

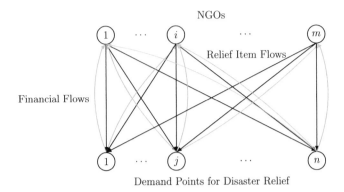

Fig. 1 The network structure of the game theory model

with these cost functions being convex and continuously differentiable. These costs also include transaction costs (see also [27]). Note that the cost functions (1) are associated with the logistics aspects and, hence, the cost on a shipment link can depend not only on its flow but also on the flows on the other shipment links associated with the same NGO or with other NGOs.

Each NGO i; $i = 1, \ldots, m$, based on the media attention and the visibility of NGOs at demand point j; $j = 1, \ldots, n$, receives financial funds from donors given by the expression

$$\sum_{j=1}^{n} P_{ij}(q), \tag{2}$$

where $P_{ij}(q)$ denotes the financial funds in donation dollars given to NGO i due to visibility of NGO i at location j. Hence, $P_{ij}(q)$ corresponds to the financial flow on the link joining demand node j with node NGO i in Figure 1. Observe that, according to (2), there is competition among all the NGOs for financial donations since the financial amount of donations that an NGO receives depends not only on its relief item deliveries but also on those delivered by other NGOs. Indeed, according to (2), an NGO may benefit from donations even through visibility of other NGOs providing the product because of, for example, loyalty and support for a specific NGO. We assume that the P_{ij} functions are increasing, concave, and continuously differentiable. Hence, we have positive but decreasing marginal utility of providing aid (in terms of the NGO's effect on attracting donations). It is important to mention that the $P_{ij}(q)$ function contains, as a special case, the financial funds donor function of [28], with $P_{ij}(q) = \beta_i P_j(q)$; $i = 1, \ldots, m$; $j = 1, \ldots, n$. Furthermore, [37] noted that the cheapest way for relief organizations to fundraise is to provide early relief in highly visible areas. In our case study in Section we construct explicit $P_{ij}(q)$ functions for all NGOs i and demand points j.

Also, since the NGOs are humanitarian organizations involved in disaster relief, each NGO i also derives some utility from delivering the needed relief supplies. We, hence, introduce an altruism/benefit function B_i; $i = 1, \ldots, m$, such that

$$B_i = B_i(q), \tag{3}$$

and each benefit function is assumed to be concave and continuously differentiable. Previously utilized benefit functions in this application domain were of the form: $B_i = \sum_{j=1}^{n} \gamma_{ij} q_{ij}$; $j = 1, \ldots, n$. Furthermore, when we construct each NGO's full utility function we will also assign a weight ω_i before each $B_i(q)$; $i = 1, \ldots, m$, to represent a monetized weight associated with altruism of i. Such weight concepts are used in multicriteria decision-making; see, e.g., [5, 14, 18, 48], and [28].

Each NGO i; $i = 1, \ldots, m$, has an amount s_i of the relief item that it can allocate post-disaster, which must satisfy:

$$\sum_{j=1}^{n} q_{ij} \leq s_i. \tag{4}$$

We assume that the relief supplies have been prepositioned so that they are in stock and available, since time is of the essence. According to [39], prepositioning of supplies can make emergency relief more effective and this is a strategy followed not only by the UNHRD (United Nations Humanitarian Response Depot) but also by the Red Cross and even some smaller relief organizations such as AmeriCares. Gatignon et al. [17] also note the benefits of proper prepositioning of supplies in the case of the International Federation of the Red Cross (IFRC) in terms of cost reduction and a more timely response.

In addition, the relief item flows for each i; $i = 1, \ldots, m$, must be nonnegative, that is:

$$q_{ij} \geq 0, \quad j = 1, \ldots, n. \tag{5}$$

Each NGO i; $i = 1, \ldots, m$, seeks to maximize its utility, U_i, with the utility consisting of the financial gains due to its visibility through media of the relief item flows, $\sum_{j=1}^{n} P_{ij}(q)$, plus the utility associated with the logistical (supply chain) aspects of delivery of the supplies, which consists of the weighted altruism/benefit function minus the logistical costs. For additional background on utility functions for nonprofit and charitable organizations, see [40] and [23].

Without the imposition of demand bound constraints (which will follow), the optimization problem faced by NGO i; $i = 1, \ldots, m$, is, thus,

$$\text{Maximize} \quad U_i(q) = \sum_{j=1}^{n} P_{ij}(q) + \omega_i B_i(q) - \sum_{j=1}^{n} c_{ij}(q) \tag{6}$$

subject to constraints (4) and (5).

Before imposing the common constraints, we remark that the above model, in the absence of any common constraints, is a Nash Equilibrium problem, which we know can be formulated and solved as a variational inequality problem (cf. [16] and [26]). Indeed, although the utility functions of the NGOs depend on their strategies and those of the other NGOs, the respective NGO feasible sets do not. However, the NGOs may be faced with several common constraints, which make the game theory problem more complex and challenging. The common constraints, which are imposed by an authority, such as a governmental one or a higher level humanitarian coordination agency, ensure that the needs of the disaster victims are met, while recognizing the negative effects of waste and material convergence. The imposition of such constraints in terms of effectiveness and even gains for NGOs was demonstrated in [28]. Later in this section, we present the variational inequality framework. Hence, we will not need to make use of quasivariational inequalities (cf. [45]) for our new model.

Specifically, the two sets of common imposed constraints, at each demand point $j; j = 1, \ldots, n$, are as follows:

$$\sum_{i=1}^{m} q_{ij} \geq \underline{d}_j, \tag{7}$$

and

$$\sum_{i=1}^{m} q_{ij} \leq \bar{d}_j, \tag{8}$$

where $\underline{d}_j$ is the lower bound on the amount of the relief item needed at demand point j and $\bar{d}_j$ is the upper bound on the amount of the relief item needed at demand point j. The constraints (7) and (8) give flexibility for a regulatory or coordinating body, since it is not likely that the demand will be precisely known in a disaster situation. It is, however, reasonable to assume that, as represented in these equations, estimates for needs assessment for the relief items will be available at the local level.

We assume that

$$\sum_{i=1}^{m} s_i \geq \sum_{j=1}^{n} \underline{d}_j. \tag{9}$$

Hence, the total supply of the relief item of the NGOs is sufficient to meet the needs at all the demand points.

We define the feasible set K_i for each NGO i as:

$$K_i \equiv \{q_i \,|\, (4) \text{ and } (5) \text{ hold}\} \tag{10}$$

and we let $K \equiv \prod_{i=1}^{m} K_i$.

In addition, we define the feasible set S consisting of the shared constraints as:

$$S \equiv \{q \mid (7) \text{ and } (8) \text{ hold}\}. \tag{11}$$

Observe that now not only does the utility of each NGO depend on the strategies, that is, the relief item flows, of the other NGOs, but so does the feasible set because of the common constraints (7) and (8). Hence, the above game theory model, in which the NGOs compete noncooperatively is a Generalized Nash Equilibrium problem. Therefore, we have the following definition.

Definition 1 (Disaster Relief Generalized Nash Equilibrium) A relief item flow pattern $q^* \in K = \prod_{i=1}^{m} K_i$, $q^* \in S$, constitutes a disaster relief Generalized Nash Equilibrium if for each NGO i; $i = 1, \ldots, m$:

$$U_i(q_i^*, \hat{q}_i^*) \geq U_i(q_i, \hat{q}_i^*), \quad \forall q_i \in K_i, \forall q \in S, \tag{12}$$

where $\hat{q}_i^* \equiv (q_1^*, \ldots, q_{i-1}^*, q_{i+1}^*, \ldots, q_m^*)$.

Hence, an equilibrium is established if no NGO can unilaterally improve upon its utility by changing its relief item flows in the disaster relief network, given the relief item flow decisions of the other NGOs, and subject to the supply constraints, the nonnegativity constraints, and the shared/coupling constraints. We remark that both K and S are convex sets.

If there are no coupling, that is, shared, constraints in the above model, then q and q^* in Definition 1 need only lie in the set K, and, under the assumption of concavity of the utility functions and that they are continuously differentiable, we know that (cf. [16] and [26]) the solution to what would then be a Nash equilibrium problem (see [35, 36]) would coincide with the solution of the following variational inequality problem: determine $q^* \in K$, such that

$$-\sum_{i=1}^{m} \langle \nabla_{q_i} U_i(q^*), q_i - q_i^* \rangle \geq 0, \quad \forall q \in K, \tag{13}$$

where $\langle \cdot, \cdot \rangle$ denotes the inner product in the corresponding Euclidean space and $\nabla_{q_i} U_i(q)$ denotes the gradient of $U_i(q)$ with respect to q_i.

As emphasized in [34], a refinement of the Generalized Nash Equilibrium is what is known as a variational equilibrium and it is a specific type of GNE (see [21]). Specifically, in a GNE defined by a variational equilibrium, the Lagrange multipliers associated with the common/shared/coupling constraints are all the same. This feature provides a fairness interpretation and is reasonable from an economic standpoint. More precisely, we have the following definition:

Definition 2 (Variational Equilibrium) A strategy vector q^* is said to be a variational equilibrium of the above Generalized Nash Equilibrium game if $q^* \in$

$K, q^* \in S$ is a solution of the variational inequality:

$$-\sum_{i=1}^{m} \langle \nabla_{q_i} U_i(q^*), q_i - q_i^* \rangle \geq 0, \quad \forall q \in K, \forall q \in S. \tag{14}$$

By utilizing a variational equilibrium, we can take advantage of the well-developed theory of variational inequalities, including algorithms (cf. [26] and the references therein), which are in a more advanced state of development and application than algorithms for quasivariational inequality problems.

We now expand the terms in variational inequality (14).

Specifically, we have that (14) is equivalent to the variational inequality: determine $q^* \in K, q^* \in S$, such that

$$\sum_{i=1}^{m} \sum_{j=1}^{n} \left[\sum_{k=1}^{n} \frac{\partial c_{ik}(q^*)}{\partial q_{ij}} - \sum_{k=1}^{n} \frac{\partial P_{ik}(q^*)}{\partial q_{ij}} - \omega_i \frac{\partial B_i(q^*)}{\partial q_{ij}} \right] \times \left[q_{ij} - q_{ij}^* \right] \geq 0,$$

$$\forall q \in K, \forall q \in S. \tag{15}$$

We now put variational inequality (15) into standard variational inequality form (see [26]), that is: determine $X^* \in \mathcal{K} \subset R^N$, such that

$$\langle F(X^*), X - X^* \rangle \geq 0, \quad \forall X \in \mathcal{K}, \tag{16}$$

where F is a given continuous function from $\mathcal{K}$ to R^N, $\mathcal{K}$ is a closed and convex set, with both the vectors $F(X)$ and X being column vectors, and $N = mn$.

We define $X \equiv q$ and $F(X)$ where component (i, j); $i = 1, \ldots, m$; $j = 1, \ldots, n$, of $F(X)$, $F_{ij}(X)$, is given by

$$F_{ij}(X) \equiv \left[\sum_{k=1}^{n} \frac{\partial c_{ik}(q)}{\partial q_{ij}} - \sum_{k=1}^{n} \frac{\partial P_{ik}(q)}{\partial q_{ij}} - \omega_i \frac{\partial B_i(q)}{\partial q_{ij}} \right] \tag{17}$$

and $\mathcal{K} \equiv K \cap S$. Then, clearly, (14) takes on the standard form (16).

Remark (Existence and Uniqueness of an Equilibrium Solution) A solution q^* of disaster relief item flows to the variational inequality problem (15) is guaranteed to exist since the function $F(X)$ in (16) is continuous under the imposed assumptions and the feasible set $\mathcal{K}$ comprised of the constraints is compact. Furthermore, it follows from the classical theory of variational inequalities (cf. [19] and [26]) that if $F(X)$ is strictly monotone, that is:

$$\langle F(X^1) - F(X^2), X^1 - X^2 \rangle > 0, \quad \forall X^1, X^2 \in \mathcal{K}, X^1 \neq X^2,$$

then the solution to the variational inequality (16) is unique, and we have a unique equilibrium product shipment pattern q^* from the NGOs to the demand points.

Lagrange Theory and Analysis of the Marginal Utilities

In this section we explore the Lagrange theory associated with variational inequality (15) and we provide an analysis of the marginal utilities at the equilibrium solution. For an application of Lagrange theory to other models, see: [8] (spatial economic models), [2] (financial networks), [43] (end-of-life products networks), [9] (random traffic networks), [4] (transplant networks), [30] (competition for blood donations).

By setting:

$$C(q) = \sum_{i=1}^{m} \sum_{j=1}^{n} \left[\sum_{k=1}^{n} \frac{\partial c_{ik}(q^*)}{\partial q_{ij}} - \sum_{k=1}^{n} \frac{\partial P_{ik}(q^*)}{\partial q_{ij}} - \omega_i \frac{\partial B_i(q^*)}{\partial q_{ij}} \right] (q_{ij} - q_{ij}^*), \quad (18)$$

variational inequality (15) can be rewritten as a minimization problem as follows:

$$\min_{\mathcal{K}} C(q) = C(q^*) = 0. \tag{19}$$

Under the previously imposed assumptions, we know that all the involved functions in (19) are continuously differentiable and convex.

We set:

$$
\begin{aligned}
a_{ij} &= -q_{ij} \leq 0, & \forall i, \ \forall j, \\
b_i &= \sum_{j=1}^{n} q_{ij} - s_i \leq 0, & \forall i, \\
c_j &= \underline{d}_j - \sum_{i=1}^{m} q_{ij} \leq 0, & \forall j, \\
e_j &= \sum_{i=1}^{m} q_{ij} - \overline{d}_j \leq 0, & \forall j,
\end{aligned}
\tag{20}
$$

and

$$\Gamma(q) = \left(a_{ij}, b_i, c_j, e_j \right)_{i=1,\dots,m;\ j=1,\dots,n}. \tag{21}$$

As a consequence, we remark that $\mathcal{K}$ can be rewritten as

$$\mathcal{K} = \{ q \in R^{mn} : \Gamma(q) \leq 0 \}. \tag{22}$$

We now consider the following Lagrange function:

$$\mathcal{L}(q, \alpha, \delta, \sigma, \varepsilon) = \sum_{i=1}^{m} \left(\sum_{j=1}^{n} c_{ij}(q) - \sum_{j=1}^{n} P_{ij}(q) - \omega_i B_i(q) \right) \tag{23}$$
$$+ \sum_{i=1}^{m} \sum_{j=1}^{n} \alpha_{ij} a_{ij} + \sum_{i=1}^{m} \delta_i b_i + \sum_{j=1}^{n} \sigma_j c_j + \sum_{j=1}^{n} \varepsilon_j e_j,$$

$$\forall q \in R_+^{mn}, \ \forall \alpha \in R_+^{mn}, \ \forall \delta \in R_+^{m}, \ \forall \sigma \in R_+^{n}, \ \forall \varepsilon \in R_+^{n},$$

where α is the vector with components: $\{\alpha_{11}, \ldots, \alpha_{mn}\}$; δ is the vector with components $\{\delta_1, \ldots, \delta_m\}$; σ is the vector with elements: $\{\sigma_1, \ldots, \sigma_n\}$, and ϵ is the vector with elements: $\{\epsilon_1, \ldots, \epsilon_n\}$.

It is easy to prove that the feasible set $\mathcal{K}$ is convex and that the Slater condition is satisfied. Then, if q^* is a minimal solution to problem (19), there exist $\alpha^* \in R_+^{mn}, \delta^* \in R_+^{m}, \sigma^* \in R_+^{n}, \varepsilon^* \in R_+^{n}$ such that the vector $(q^*, \alpha^*, \delta^*, \sigma^*, \varepsilon^*)$ is a saddle point of the Lagrange function (23); namely:

$$\mathcal{L}(q^*, \alpha, \delta, \sigma, \varepsilon) \leq \mathcal{L}(q^*, \alpha^*, \delta^*, \sigma^*, \varepsilon^*) \leq \mathcal{L}(q, \alpha^*, \delta^*, \sigma^*, \varepsilon^*), \tag{24}$$

$$\forall q \in R_+^{mn}, \ \forall \alpha \in R_+^{mn}, \ \forall \delta \in R_+^{m}, \ \forall \sigma \in R_+^{n}, \ \forall \varepsilon \in R_+^{n},$$

and

$$\alpha_{ij}^* a_{ij}^* = 0, \ \forall i, \forall j,$$

$$\delta_i^* b_i^* = 0, \ \forall i,$$

$$\sigma_j^* c_j^* = 0, \quad \varepsilon_j^* e_j^* = 0, \ \forall j. \tag{25}$$

From the right-hand side of (24), it follows that $q^* \in R_+^{mn}$ is a minimal point of $\mathcal{L}(q, \alpha^*, \delta^*, \sigma^*, \varepsilon^*)$ in the whole space R^{mn}, and hence, for all $i = 1, \ldots, m$, and for all $j = 1, \ldots, n$, we have that:

$$\frac{\partial \mathcal{L}(q^*, \alpha^*, \delta^*, \sigma^*, \varepsilon^*)}{\partial q_{ij}}$$

$$= \sum_{k=1}^{n} \frac{\partial c_{ik}(q^*)}{\partial q_{ij}} - \sum_{k=1}^{n} \frac{\partial P_{ik}(q^*)}{\partial q_{ij}} - \omega_i \frac{\partial B_i(q^*)}{\partial q_{ij}} - \alpha_{ij}^* + \delta_i^* - \sigma_j^* + \varepsilon_j^* = 0, \tag{26}$$

together with conditions (25).

Conditions (25) and (26) represent an equivalent formulation of variational inequality (15). Indeed, if we multiply (26) by $(q_{ij} - q_{ij}^*)$ and sum up with respect to i and j, we get:

$$
\sum_{i=1}^{m}\sum_{j=1}^{n}\left[\sum_{k=1}^{n}\frac{\partial c_{ik}(q^*)}{\partial q_{ij}}-\sum_{k=1}^{n}\frac{\partial P_{ik}(q^*)}{\partial q_{ij}}-\omega_i\frac{\partial B_i(q^*)}{\partial q_{ij}}\right](q_{ij}-q_{ij}^*)
$$

$$
=\sum_{i=1}^{m}\sum_{j=1}^{n}\alpha_{ij}^*q_{ij}-\underbrace{\sum_{i=1}^{m}\sum_{j=1}^{n}\alpha_{ij}^*q_{ij}^*}_{=0}-\sum_{i=1}^{m}\left(\delta_i^*\sum_{j=1}^{n}q_{ij}-\underbrace{\delta_i^*\sum_{j=1}^{n}q_{ij}^*}_{=\delta_i^*s_i}\right)
$$

$$
+\sum_{j=1}^{n}\left(\sigma_j^*\sum_{i=1}^{m}q_{ij}-\underbrace{\sigma_j^*\sum_{i=1}^{m}q_{ij}^*}_{=\sigma_j^*\underline{d}_j}\right)-\sum_{j=1}^{n}\left(\varepsilon_j^*\sum_{i=1}^{m}q_{ij}-\underbrace{\varepsilon_j^*\sum_{i=1}^{m}q_{ij}^*}_{=\varepsilon_j^*\overline{d}_j}\right)
$$

$$
=\sum_{i=1}^{m}\sum_{j=1}^{n}\underbrace{\alpha_{ij}^*}_{\geq0}q_{ij}-\sum_{i=1}^{m}\delta_i^*\underbrace{\left(\sum_{j=1}^{n}q_{ij}-s_i\right)}_{\leq0}+\sum_{j=1}^{n}\sigma_j^*\underbrace{\left(\sum_{i=1}^{m}q_{ij}-\underline{d}_j\right)}_{\geq0}
$$

$$
-\sum_{j=1}^{n}\varepsilon_j^*\underbrace{\left(\sum_{i=1}^{m}q_{ij}-\overline{d}_j\right)}_{\leq0}\geq0. \tag{27}
$$

We now discuss the meaning of some of the Lagrange multipliers. We focus on the case where $q_{ij}^*>0$; namely, the relief item flow from NGO i to demand point j is positive; otherwise, if $q_{ij}^*=0$, the problem is not interesting. Then, from the first line in (25), we have that $\alpha_{ij}^*=0$.

Let us consider the situation when the constraints are not active, that is, $b_i^*<0$ and $\underline{d}_j<\sum_{i=1}^{m}q_{ij}^*<\overline{d}_j$.

Specifically, $b_i^*<0$ means that $\sum_{j=1}^{n}q_{ij}^*<s_i$; that is, the sum of relief items sent by the i-th NGO to all demand points is strictly less than the total amount s_i at its disposal. Then, from the second line in (25), we get: $\delta_i^*=0$.

At the same time, from the last line in (25), $\underline{d}_j < \sum\limits_{i=1}^{m} q_{ij}^* < \overline{d}_j$, leads to: $\sigma_j^* = \varepsilon_j^* = 0$.

Hence, (26) yields:

$$\sum_{k=1}^{n} \frac{\partial c_{ik}(q^*)}{\partial q_{ij}} - \sum_{k=1}^{n} \frac{\partial P_{ik}(q^*)}{\partial q_{ij}} - \omega_i \frac{\partial B_i(q^*)}{\partial q_{ij}} = \alpha_{ij}^* - \delta_i^* + \sigma_j^* - \varepsilon_j^* = 0$$

$$\Longleftrightarrow \sum_{k=1}^{n} \frac{\partial P_{ik}(q^*)}{\partial q_{ij}} + \omega_i \frac{\partial B_i(q^*)}{\partial q_{ij}} = \sum_{k=1}^{n} \frac{\partial c_{ik}(q^*)}{\partial q_{ij}}. \tag{28}$$

In this case, the marginal utility associated with the financial donations plus altruism is equal to the marginal costs.

If, on the other hand, $\sum\limits_{i=1}^{m} q_{ij}^* = \underline{d}_j$, then $\sigma_j^* > 0$. Hence, we get:

$$\sum_{k=1}^{n} \frac{\partial P_{ik}(q^*)}{\partial q_{ij}} + \omega_i \frac{\partial B_i(q^*)}{\partial q_{ij}} + \sigma_j^* = \sum_{k=1}^{n} \frac{\partial c_{ik}(q^*)}{\partial q_{ij}}, \quad \text{with } \sigma_j^* > 0, \tag{29}$$

and, therefore,

$$\sum_{k=1}^{n} \frac{\partial c_{ik}(q^*)}{\partial q_{ij}} > \sum_{k=1}^{n} \frac{\partial P_{ik}(q^*)}{\partial q_{ij}} + \omega_i \frac{\partial B_i(q^*)}{\partial q_{ij}}, \tag{30}$$

which means that the marginal costs are greater than the marginal utility associated with the financial donations plus altruism and this is a very bad situation.

Finally, if $\sum\limits_{i=1}^{m} q_{ij}^* = \overline{d}_j$, then $\varepsilon_j^* > 0$, we have that:

$$\sum_{k=1}^{n} \frac{\partial P_{ik}(q^*)}{\partial q_{ij}} + \omega_i \frac{\partial B_i(q^*)}{\partial q_{ij}} = \sum_{k=1}^{n} \frac{\partial c_{ik}(q^*)}{\partial q_{ij}} + \varepsilon_j^*, \quad \text{with } \varepsilon_j^* > 0. \tag{31}$$

Therefore,

$$\sum_{k=1}^{n} \frac{\partial c_{ik}(q^*)}{\partial q_{ij}} < \sum_{k=1}^{n} \frac{\partial P_{ik}(q^*)}{\partial q_{ij}} + \omega_i \frac{\partial B_i(q^*)}{\partial q_{ij}}. \tag{32}$$

In this situation, the relevant marginal utility exceeds the marginal cost and this is a desirable situation.

Analogously, if we assume that the conservation of flow equation is active; that is, if $\sum_{j=1}^{n} q_{ij}^* = s_i$, then $\delta_i^* > 0$. As a consequence, we obtain:

$$\sum_{k=1}^{n} \frac{\partial P_{ik}(q^*)}{\partial q_{ij}} + \omega_i \frac{\partial B_i(q^*)}{\partial q_{ij}} = \sum_{k=1}^{n} \frac{\partial c_{ik}(q^*)}{\partial q_{ij}} + \delta_i^*, \text{ with } \delta_i^* > 0, \quad (33)$$

which means that, once again, the marginal utility associated with the financial donations plus altruism exceeds the marginal cost and this is the desirable situation.

From the above analysis of the Lagrange multipliers and marginal utilities at the equilibrium solution, we can conclude that the most convenient situation, in terms of the marginal utilities, is the one when $\sum_{i=i}^{m} q_{ij}^* = \overline{d}_j$ and $\sum_{j=1}^{n} q_{ij}^* = s_i$.

Taking into account the Lagrange multipliers, an equivalent variational formulation of problem (6) under constraints (4), (5), (7), and (8) is the following one:

$$\text{Find } (q^*, \delta^*, \sigma^*, \varepsilon^*) \in R_+^{mn+m+2n}:$$

$$\sum_{i=1}^{m} \sum_{j=1}^{n} \left[\sum_{k=1}^{n} \frac{\partial c_{ik}(q^*)}{\partial q_{ij}} - \sum_{k=1}^{n} \frac{\partial P_{ik}(q^*)}{\partial q_{ij}} - \omega_i \frac{\partial B_i(q^*)}{\partial q_{ij}} + \delta_i^* - \sigma_j^* + \varepsilon_j^* \right] (q_{ij} - q_{ij}^*)$$

$$+ \sum_{i=1}^{m} \left(s_i - \sum_{j=1}^{n} q_{ij}^* \right) (\delta_i - \delta_i^*)$$

$$+ \sum_{j=1}^{n} \left(\sum_{i=1}^{m} q_{ij}^* - \underline{d}_j \right) (\sigma_j - \sigma_j^*) + \sum_{j=1}^{n} \left(\overline{d}_j - \sum_{i=1}^{m} q_{ij}^* \right) (\varepsilon_j - \varepsilon_j^*) \geq 0,$$

$$(34)$$

$$\forall q \in R_+^{mn}, \ \forall \delta \in R_+^{m}, \ \forall \sigma \in R_+^{n}, \ \forall \varepsilon \in R_+^{n}.$$

The Algorithm and Case Study

Before we present the case study, we outline the algorithm that we utilize for the computations, notably, the Euler method of [11], since it nicely exploits the feasible set underlying variational inequality (34), which is simply the nonnegative orthant.

Recall that, as established in [11], for convergence of the general iterative scheme, which induces the Euler method, the sequence $\{a_\tau\}$ must satisfy: $\sum_{\tau=0}^{\infty} a_\tau = \infty, a_\tau > 0, a_\tau \to 0$, as $\tau \to \infty$. Conditions for convergence for a variety of network-based problems can be found in [32] and [27].

Specifically, at iteration τ, the Euler method yields the following closed form expressions for the relief item flows and the Lagrange multipliers.

Explicit Formulae for the Euler Method Applied to the Game Theory Model

In particular, we have the following closed form expression for the relief item flows $i = 1, \ldots, m; j = 1, \ldots, n$, at each iteration $\tau + 1$:

$$q_{ij}^{\tau+1} = \max\{0, q_{ij}^{\tau} + a_{\tau}(\sum_{k=1}^{n} \frac{\partial P_{ik}(q^{\tau})}{\partial q_{ij}} + \omega_i \frac{\partial B_i(q^{\tau})}{\partial q_{ij}} - \sum_{k=1}^{n} \frac{\partial c_{ik}(q^{\tau})}{\partial q_{ij}} - \delta_i^{\tau} + \sigma_j^{\tau} - \epsilon_j^{\tau})\};$$

(35)

the following closed form expressions for the Lagrange multipliers associated with the supply constraints (4), respectively, for $i = 1, \ldots, m$:

$$\delta_i^{\tau+1} = \max\{0, \delta_i^{\tau} + a_{\tau}(-s_i + \sum_{j=1}^{n} q_{ij}^{\tau})\};$$

(36)

the following closed form expressions for the Lagrange multipliers associated with the lower bound demand constraints (7), respectively, for $j = 1, \ldots, n$:

$$\sigma_j^{\tau+1} = \max\{0, \sigma_j^{\tau} + a_{\tau}(-\sum_{i=1}^{m} q_{ij}^{\tau} + \underline{d}_j)\},$$

(37)

and the following closed form expressions for the Lagrange multipliers associated with the upper bound demand constraints (8), respectively, for $j = 1, \ldots, n$:

$$\epsilon_j^{\tau+1} = \max\{0, \epsilon_j^{\tau} + a_{\tau}(-\bar{d}_j + \sum_{i=1}^{m} q_{ij}^{\tau})\}.$$

(38)

Our case study is inspired by a disaster consisting of a series of tornados that hit western Massachusetts on June 1, 2011 in the late afternoon. The largest tornado was measured at EF3. It was the worst tornado outbreak in the area in a century (see [15]). A wide swath from western to central Massachusetts was impacted. According to the [47]: "The tornado caused extensive damage, killed 4 persons, injured more than 200 persons, damaged or destroyed 1500 homes, left over 350 people homeless in Springfield's MassMutual Center arena, left 50,000 customers without power, and brought down thousands of trees." The same report notes that: FEMA estimated that 1435 residences were impacted with the following breakdowns: 319 destroyed, 593 sustaining major damage, 273 sustaining minor damage, and 250 otherwise affected. FEMA estimated that the primary impact was damage to buildings and equipment with a cost estimate of $24,782,299. Total damage estimates from the storm exceeded $140 million, the majority from the destruction of homes and businesses.

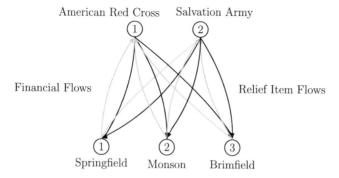

American Red Cross   Salvation Army

Financial Flows   Relief Item Flows

Springfield   Monson   Brimfield

Fig. 2 The network topology for the case study, Example 1

Especially impacted were the city of Springfield and the towns of Monson and Brimfield. It has been estimated that in the aftermath, the Red Cross served about 11,800 meals and the Salvation Army about 20,000 meals (cf. [47]).

The network topology for our case study, Example 1, is depicted in Figure 2. The NGO nodes consist of the American Red Cross and the Salvation Army, respectively. The demand points correspond to Springfield, Monson, and Brimfield, respectively.

Example 1 The data for our case study, Example 1, are given below. The supplies of meals available for delivery to the victims are:

$$s_1 = 25,000, \quad s_2 = 25,000,$$

with the weights associated with the altruism benefit functions of the NGOs given by:

$$\omega_1 = 1, \quad \omega_2 = 1.$$

The financial funds functions are:

$$P_{11}(q) = 1000\sqrt{(3q_{11} + q_{21})}, \quad P_{12}(q) = 600\sqrt{(2q_{12} + q_{22})},$$
$$P_{13}(q) = 400\sqrt{(2q_{13} + q_{23})},$$

$$P_{21}(q) = 800\sqrt{(4q_{21} + q_{11})}, \quad P_{22}(q) = 400\sqrt{(2q_{22} + q_{12})},$$
$$P_{23}(q) = 200\sqrt{(2q_{23} + q_{13})}.$$

The altruism functions are:

$$B_1(q) = 300q_{11} + 200q_{12} + 100q_{13}, \quad B_2(q) = 400q_{21} + 300q_{22} + 200q_{23}.$$

The cost functions, which capture distance from the main storage depots in Springfield, are:

$$c_{11}(q) = .15q_{11}^2 + 2q_{11}, \quad c_{12}(q) = .15q_{12}^2 + 5q_{12}, \quad c_{13}(q) = .15q_{13}^2 + 7q_{13},$$

$$c_{21}(q) = .1q_{21}^2 + 2q_{21}, \quad c_{22}(q) = .1q_{22}^2 + 5q_{22}, \quad c_{23}(q) = .1q_{23}^2 + 7q_{23}.$$

The demand lower and upper bounds at the three demand points are:

$$\underline{d}_1 = 10000, \quad \bar{d}_1 = 20000,$$

$$\underline{d}_2 = 1000, \quad \bar{d}_2 = 10000,$$

$$\underline{d}_3 = 1000, \quad \bar{d}_3 = 10000.$$

The Euler method was implemented in FORTRAN and a Linux system at the University of Massachusetts Amherst was used for the computations. The algorithm was initialized as follows: all Lagrange multipliers were set to 0.00 and the initial relief item flows to a given demand point were set to the lower bound divided by the number of NGOs, which here is two.

The Euler method yielded the following Generalized Nash Equilibrium solution: The equilibrium relief item flows are:

$$q_{11}^* = 3800.24, \quad q_{12}^* = 668.64, \quad q_{13}^* = 326.66,$$

$$q_{21}^* = 6199.59, \quad q_{22}^* = 1490.52, \quad q_{23}^* = 974.97.$$

Since none of the supplies are exhausted, the computed Lagrange multipliers associated with the supply constraints are:

$$\delta_1^* = 0.00, \quad \delta_2^* = 0.00.$$

Since the demand at the first demand point, which is the city of Springfield, is essentially at its lower bound, we have that:

$$\sigma_1^* = 835.22,$$

with

$$\sigma_2^* = 0.00, \quad \sigma_3^* = 0.00.$$

All the Lagrange multipliers associated with the demand upper bound constraints are equal to zero, that is:

$$\epsilon_1^* = \epsilon_2^* = \epsilon_3^* = 0.00.$$

In terms of the gain in financial donations, the NGOs receive the following amounts:

$$\sum_{j=1}^{3} P_{1j}(q^*) = 180,713.23, \quad \sum_{j=1}^{3} P_{2j}(q^*) = 168,996.78.$$

This is reasonable since the American Red Cross tends to have greater visibility post disasters than the Salvation Army through the media and that was the case post the Springfield tornadoes.

We then proceeded to solve the Nash equilibrium counterpart of the above Generalized Nash Equilibrium problem formulated as a variational equilibrium. The variational inequality for the Nash equilibrium is given in (13) and does not include the upper and lower bound demand constraints. We solved it using the Euler method but over the feasible set K as in (13).

The computed equilibrium relief item flows for the Nash equilibrium are:

$$q_{11}^* = 1040.22, \quad q_{12}^* = 668.64, \quad q_{13}^* = 326.66,$$

$$q_{21}^* = 2054.51, \quad q_{22}^* = 1490.52, \quad q_{23}^* = 974.97.$$

The Lagrange multipliers associated with the supply constraints are:

$$\delta_1^* = 0.00, \quad \delta_2^* = 0.00.$$

Observe that, without the imposition of the bounds on the demands, Springfield, which is demand point 1 and is a big city, receives only about one third of the volume of supplies (in this case, meals) as needed, and as determined by the Generalized Nash equilibrium solution.

The American Red Cross now garners financial donations of: 119,985.66, whereas the Salvation Army stands to receive financial donations equal to: 110,683.60. These values are significantly lower than the analogous ones for the Generalized Nash equilibrium model above. Hence, NGOs, by coordinating their deliveries of needed supplies, such as meals, can gain in terms of financial donations and attend to the victims' needs better by delivering in the amounts that have been estimated to be needed in terms of lower and upper bounds. This more general model, for which an optimization reformulation does not exist, in contrast to the model of [33], nevertheless, supports the numerical result findings in the case study for Katrina therein.

Example 2 We now investigate the possible impact of the addition of a new disaster relief organization, such as a church-based one, or the Springfield Partners for Community Action, which also assisted in disaster relief, providing meals post the tornadoes. Hence, the network topology for case study, Example 2, is as in Figure 3. We refer to the added NGO as "Other." It is based in Springfield.

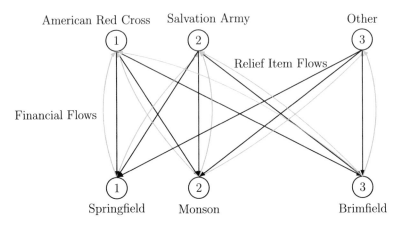

Fig. 3 The network topology for the case study, Example 2

The data are as in Example 1 but with the original $P_{ij}(q)$ functions for the America Red Cross and the Salvation Army expanded as per below and the added data for the "Other" NGO also as given below.

The financial funds functions are now:

$$P_{11}(q) = 1000\sqrt{(3q_{11} + q_{21} + q_{31})}, \quad P_{12}(q) = 600\sqrt{(2q_{12} + q_{22} + q_{32})},$$
$$P_{13}(q) = 400\sqrt{(2q_{13} + q_{23} + q_{33})},$$

$$P_{21}(q) = 800\sqrt{(4q_{21} + q_{11} + q_{31})}, \quad P_{22}(q) = 400\sqrt{(2q_{22} + q_{12} + q_{32})},$$
$$P_{23}(q) = 200\sqrt{(2q_{23} + q_{13} + q_{33})},$$

with those for the new NGO:

$$P_{31}(q) = 400\sqrt{(2q_{31} + q_{11} + q_{21})}, \quad P_{32}(q) = 200\sqrt{(2q_{32} + q_{12} + q_{22})},$$
$$P_{33}(q) = 100\sqrt{(2q_{33} + q_{13} + q_{23})}.$$

The weight $\omega_3 = 1$ and the altruism/benefit function for the new NGO is:

$$B_3(q) = 200q_{31} + 100q_{32} + 100q_{33}.$$

The cost functions associated with the added NGO are:

$$c_{31}(q) = .1q_{31}^2 + q_{31}, \quad c_{32}(q) = .2q_{32}^2 + 5q_{32}, \quad c_{33}(q) = .2q_{33}^2 + 7q_{33}.$$

The Euler method converged to the following Generalized Nash Equilibrium solution:

The equilibrium relief item flows are:

$$q_{11}^* = 2506.97, \quad q_{12}^* = 667.85, \quad q_{13}^* = 325.59,$$

$$q_{21}^* = 4259.59, \quad q_{22}^* = 1489.98, \quad q_{23}^* = 974.45,$$

$$q_{31}^* = 3233.35, \quad q_{32}^* = 242.42, \quad q_{33}^* = 235.52.$$

Since none of the supplies are exhausted, the computed Lagrange multipliers associated with the supply constraints are:

$$\delta_1^* = 0.00, \quad \delta_2^* = 0.00, \quad \delta_3^* = 0.00.$$

The demand at the first demand point, which is the city of Springfield, is at the lower bound of 10000.00. Hence, we have that:

$$\sigma_1^* = 446.70,$$

with

$$\sigma_2^* = 0.00, \quad \sigma_3^* = 0.00.$$

All the Lagrange multipliers associated with the demand upper bound constraints are equal to zero, that is:

$$\epsilon_1^* = \epsilon_2^* = \epsilon_3^* = 0.00.$$

In terms of the gain in financial donations, the NGOs receive the following amounts:

$$\sum_{j=1}^{3} P_{1j}(q^*) = 173,021.70, \quad \sum_{j=1}^{3} P_{2j}(q^*) = 155,709.50,$$

$$\sum_{j=1}^{3} P_{3j}(q^*) = 60,504.14.$$

The volumes of relief items from the American Red Cross and the Salvation Army to Springfield are greatly reduced, as compared to the respective volumes in Example 1 and both original NGOs in Example 1 now experience a reduction in financial donations because of the increased competition for financial donations.

For completeness, we also solved the Nash equilibrium counterpart for Example 2.

The Nash equilibrium relief item flows are:

$$q_{11}^* = 1036.27, \quad q_{12}^* = 667.85, \quad q_{13}^* = 325.59,$$

$$q_{21}^* = 2051.17, \quad q_{22}^* = 1489.98, \quad q_{23}^* = 974.45,$$

$$q_{31}^* = 1009.61, \quad q_{32}^* = 242.42, \quad q_{33}^* = 235.52.$$

The financial donations of the NGOs are now the following:

$$\sum_{j=1}^{3} P_{1j}(q^*) = 129,037.42, \quad \sum_{j=1}^{3} P_{2j}(q^*) = 115,964.80, \quad \sum_{j=1}^{3} P_{3j}(q^*) = 43,07.16.$$

In Example 2 of our case study, we, again, see that the NGOs garner greater financial funds through the Generalized Nash Equilibrium solution, rather than the Nash equilibrium one. Moreover, the needs of the victims are met under the Generalized Nash Equilibrium solution. Hence, both the NGOs and victims gain from enhanced coordination, as provided by the relief item demand upper and lower bounds at the demand points.

Summary and Conclusions

In this paper, we constructed a new Generalized Nash Equilibrium (GNE) model for disaster relief, which contains both logistical and financial funds aspects. The NGOs compete for financial funds through their visibility in the response to a disaster and provide needed supplies to the victims. A coordinating body imposes upper bounds and lower bounds for the supplies at the various demand points to guarantee that the victims receive the amounts at the points of demand that are needed, and without excesses that can add to the congestion and materiel convergence. The model is more general than the one proposed earlier by Nagurney et al. [33] and no longer is it possible to reformulate the governing equilibrium conditions as an optimization problem.

Here we use a variational equilibrium formulation of the Generalized Nash Equilibrium, which is then amenable to solution via variational inequality algorithms. We provide qualitative properties of the equilibrium pattern and also utilize Lagrange theory for the analysis of the NGOs' marginal utilities.

The proposed computational scheme yields closed form expressions, at each iteration, for the product flows and the Lagrange multipliers. The algorithm is then applied to a case study, inspired by rare tornadoes that caused devastation in parts of western and central Massachusetts in 2011. For completeness, we also compute the solution to the Nash equilibrium counterparts of the two examples making up the case study, in which the common demand bound constraints are removed. The case study reveals that victims may not receive the required amounts of supplies, without the imposition of the demand bounds. These results provide further support for the need for greater coordination in disaster relief. Moreover, by delivering the required amounts of supplies the NGOs can also garner greater financial donations.

The framework constructed here presents many opportunities for developing extensions and enhancements such as:

1. Allowing for the purchase of additional relief supplies, which we have assumed to be prepositioned and to be sufficient to meet the minimal demands, with associated costs;
2. Expanding the logistical side to include detailed transportation networks with mode options;
3. Constructing a multiperiod version of the model in which the financial flows that are not earmarked for a specific disaster can be transferred for use in other disasters as well as for the day to day operations of the NGOs;
4. Including stochastic components associated with costs, financial donations, and/or the demands for relief items at the demand points, and, finally,
5. Capturing the potential for cooperation, rather than noncooperative behavior, especially in terms of sharing resources for relief item deliveries to demand points.

We leave the above for future research. Furthermore, we emphasize that disseminating the results to practitioners is also essential and that includes emphasizing what game theory can offer in terms of disaster relief as done in [29].

Acknowledgements This paper is dedicated to the memory of Professor Martin J. Beckmann, Professor Emeritus at Brown University, who passed away on April 11, 2017 at the age of 92. He was a renowned scholar in transportation science, regional science, and operations research, and his work on network equilibria has had a profound impact on both theory and practice.

The first author acknowledges support from the Radcliffe Institute for Advanced Study at Harvard University where she was a 2017 Summer Fellow. The second author acknowledges support from the project PON SCN_00451 CLARA—CLoud plAtform and smart underground imaging for natural Risk Assessment, Smart Cities and Communities and Social Innovation.

The authors acknowledge the co-organizers of the 3rd Dynamics of Disasters conference as well as the participants for a stimulating conference and discussions. They also thank the anonymous reviewer for helpful comments and suggestions on an earlier version of this paper.

References

1. Balcik, B., Beamon, B., Krejci, C., Muramatsu, K.M., Ramirez, M.: Coordination in humanitarian relief chains: practices, challenges and opportunities. Int. J. Prod. Econ. **120**(1), 22–34 (2010)
2. Barbagallo, A., Daniele, P., Maugeri, A.: Variational formulation for a general dynamic nancial equilibrium problem: balance law and liability formula. Nonlinear Anal. Theory Methods Appl. **75**(3), 1104–1123 (2012)
3. Bensoussan, A.: Points de Nash dans le cas de fonctionelles quadratiques et jeux differentiels lineaires a N personnes. SIAM J. Control **12**, 460–499 (1974)

4. Caruso, V., Daniele, P.: A network model for minimizing the total organ transplant costs. Eur. J. Oper. Res. **266**(2), 652–662 (2018)
5. Chankong, V., Haimes, Y.Y.: Multiobjective Decision Making: Theory and Methodology. North-Holland, New York (1983)
6. Coles, J.B., Zhuang, J.: Decisions in disaster recovery operations: a game theoretic perspective on organization cooperation. J. Homeland Secur. Emerg. Manag. **8**(1), Article 35 (2011)
7. Cox, C.W.: Manmade disasters: a historical review of terrorism and implications for the future. Online J. Nurs. **13**(1). 8 Jan 2008. Available at: http://www.nursingworld.org/MainMenuCategories/ANAMarketplace/ANAPeriodicals/OJIN/TableofContents/vol132008/No1Jan08/ArticlePreviousTopic/ManmadeDisasters.html
8. Daniele, P.: Variational inequalities for static equilibrium market. In: Giannessi, F., Maugeri, A., Pardalos, P.M. (eds.) Lagrangean Function and Duality, in Equilibrium Problems: Nonsmooth Optimization and Variational Inequality Models, pp. 43–58. Kluwer Academic Publishers, Norwell (2001)
9. Daniele, P., Giuffrè, S.: Random variational inequalities and the random traffic equilibrium problem. J. Optim. Theory Appl. **167**(1), 363–381 (2015)
10. Debreu, G.: A social equilibrium existence theorem. Proc. Natl. Acad. Sci. U. S. A. **38**, 886–893 (1952)
11. Dupuis, P., Nagurney, A.: Dynamical systems and variational inequalities. Ann. Oper. Res. **44**, 9–42 (1993)
12. Facchinei, F., Kanzow, C.: Generalized Nash equilibrium problems. Ann. Oper. Res. **175**(1), 177–211 (2010)
13. Fischer, A., Herrich, M., Schonefeld, K.: Generalized Nash equilibrium problems – recent advances and challenges. Pesquisa Operacional **34**(3), 521–558 (2014)
14. Fishburn, P.C.: Utility Theory for Decision Making. Wiley, New York (1970)
15. Flynn, J.: Tornado tears through Pioneer Valley, killing two, damaging homes in 9 communities and causing widespread power outages. 2 Jun 2011. Available at: http://www.masslive.com/news/index.ssf/2011/06/tornado_tears_through_pioneer.html
16. Gabay, D., Moulin, H.: On the uniqueness and stability of Nash equilibria in noncooperative games. In: Bensoussan, A., Kleindorfer, P., Tapiero, C.S. (eds.) Applied Stochastic Control of Econometrics and Management Science, pp. 271–294. North-Holland, Amsterdam (1980)
17. Gatignon, A., Van Wassenhove, L., Charles, A.: The Yogyakarta earthquake: humanitarian relief through IFRC's decentralized supply chain. Int. J. Prod. Econ. **126**(1), 102–110 (2010)
18. Keeney, R.L., Raiffa, H.: Decisions with Multiple Objectives: Preferences and Value Tradeoffs. Cambridge University Press, Cambridge (1993)
19. Kinderlehrer, D., Stampacchia, G.: Variational Inequalities and Their Applications. Academic Press, New York (1980)
20. Kopinak, J.K.: Humanitarian assistance: are effective and sustainability impossible dreams? J. Humanitarian Assistance. March 10 (2013)
21. Kulkarni, A.A., Shanbhag, U.V.: On the variational equilibrium as a refinement of the generalized Nash equilibrium. Automatica **48**, 45–55 (2012)
22. Luna, J.P.: Decomposition and approximation methods for variational inequalities, with applications to deterministic and stochastic energy markets. PhD thesis, Instituto Nacional de Matematica Pura e Aplicada, Rio de Janeiro, Brazil (2013)
23. Malani, A., Philipson, T., David, G.: Theories of firm behavior in the nonprofit sector. A synthesis and empirical evaluation. In: Glaeser, E.L. (ed.) The Governance of Not-for-Profit Organizations, pp. 181–215. University of Chicago Press, Chicago (2003)
24. Muggy, L.: Quantifying and Mitigating Decentralized Decision Making in Humanitarian Logistics Systems. PhD Dissertation, Kansas State University, Manhattan (2015)
25. Muggy L., Heier Stamm, J.L.: Game theory applications in humanitarian operations: a review. J. Humanitarian Logist. Supply Chain Manag. **4**(1), 4–23 (2014)
26. Nagurney, A.: Network Economics: A Variational Inequality Approach, 2nd rev. ed. Kluwer Academic Publishers, Dordrecht (1999)

27. Nagurney, A.: Supply Chain Network Economics: Dynamics of Prices, Flows, and Profits. Edward Elgar Publishing, Cheltenham (2006)
28. Nagurney, A.: Freight service provision for disaster relief: A competitive network model with computations. In: Kotsireas, I.S., Nagurney, A., Pardalos, P.M. (eds.) Dynamics of Disasters: Key Concepts, Models, Algorithms, and Insights, pp. 207–229. Springer International Publishing, Switzerland (2016)
29. Nagurney, A.: Response to natural disasters like Harvey could be helped with game theory. The Conversation, August 28 (2017)
30. Nagurney, A., Dutta, P.: Competition for Blood Donations: A Nash Equilibrium Network Framework. Isenberg School of Management, University of Massachusetts Amherst (2017)
31. Nagurney, A., Qiang, Q.: Fragile Networks: Identifying Vulnerabilities and Synergies in an Uncertain World. Wiley, Hoboken (2009)
32. Nagurney, A., Zhang, D.: Projected Dynamical Systems and Variational Inequalities with Applications. Kluwer Academic Publishers, Boston (1996)
33. Nagurney, A., Alvarez Flores, E., Soylu, C.: A Generalized Nash Equilibrium model for post-disaster humanitarian relief. Transp. Res. E **95**, 1–18 (2016)
34. Nagurney, A., Yu, M., Besik, D.: Supply chain network capacity competition with outsourcing: a variational equilibrium framework. J. Glob. Optim. **69**(1), 231–254 (2017)
35. Nash, J.F.: Equilibrium points in n-person games. Proc. Natl. Acad. Sci. U. S. A. **36**, 48–49 (1950)
36. Nash, J.F.: Noncooperative games. Ann. Math. **54**, 286–298 (1951)
37. Natsios, A.S.: NGOs and the UN system in complex humanitarian emergencies: conflict or cooperation? Third World Q. **16**(3), 405419 (1995)
38. NOAA: Billion-dollar weather and climate disasters: overview (2017). Available at: https://www.ncdc.noaa.gov/billions/
39. Roopanarine, L.: How pre-positioning can make emergency relief more effective. The Guardian 17 Jan 2013
40. Rose-Ackerman, S.: Charitable giving and "excessive" fundraising. Q. J. Econ. **97**(2), 193–212 (1982)
41. Sumpf, D., Isaila, V., Najjar, K.: The impact of the Syria crisis on Lebanon. In: Kotsireas, I.S., Nagurney, A., Pardalos, P.M. (eds.) Dynamics of Disasters: Key Concepts, Models, Algorithms, and Insights, pp. 269–308. Springer International Publishing, Switzerland (2016)
42. Toyasaki, F., Wakolbinger, T.: Impacts of earmarked private donations for disaster fundraising. Ann. Oper. Res. **221**, 427–447 (2014)
43. Toyasaki, F., Daniele, P., Wakolbinger, T.: A variational inequality formulation of equilibrium models for end-of-life products with nonlinear constraints. Eur. J. Oper. Res. **236**(1), 340–350 (2014)
44. Van Wassenhove, L.N.: Blackett memorial lecture. Humanitarian aid logistics: supply chain management in high gear. J. Oper. Res. Soc. **57**(5), 475–489 (2006)
45. von Heusinger, A.: Numerical Methods for the Solution of the Generalized Nash Equilibrium Problem. PhD Dissertation, University of Wurtburg, Germany (2009)
46. Watson, C., Caravani, A., Mitchell, T., Kellett, J., Peters, K.: 10 things to know about finance for reducing disaster risk, Overseas Development Institute, London, England (2015)
47. Western Massachusetts Regional Homeland Security Advisory Council: June 1, 2011 tornado response: After action report and improvement plan (2012). Available at: http://www.wrhsac.org/frcog/Mass%20Tornado%20AAR%20IP%20Jan%202012.pdf
48. Yu, P.L.: Multiple Criteria Decision Making Concepts, Techniques, and Extensions. Plenum Press, New York (1985)
49. Zhuang, J., Saxton, G., Wu, H.: Publicity vs. impact in nonprofit disclosures and donor preferences: A sequential game with one nonprofit organization and N donors. Ann. Oper. Res. **221**(1), 469–491 (2014)

Advances in Disaster Communications: Broadband Systems for First Responders

Ladimer S. Nagurney

Abstract First responders depend upon reliable communications to function effectively. During the past eight decades, their communications needs have expanded from simple two-way voice radios to highly integrated voice and data transmissions that allow not only command messages to be transmitted, but also data and images to be shared. These days, the need for data at the site of a disaster requires responders to be able to access broadband networks that are connected to the internet. Simultaneously, the communications of the general public have evolved from the utilization of individual landlines to the ubiquitous smart phones of today. The broadband communications needs of first responders and the general public utilize the same broadband networks. Hence, when a disaster occurs, these broadband networks become overloaded and communications and access become unreliable. To rectify this problem for first responders, a dedicated First Responder Network, FirstNet, has been created as a public-private partnership that will allow priority access for first responders at a disaster scene. In this article, I begin by outlining the evolution of communications technology that led to the development of FirstNet. I then describe the political process that supported the establishment of FirstNet, the implementation of this public-private partnership, and the awarding of the FirstNet contract to a consortium led by AT&T. I highlight complementary technologies to those used in FirstNet and illustrate how they may be used in concert with FirstNet's LTE network to further extend communications opportunities for first responders. I conclude with a discussion of some possible enhancements to the LTE technology chosen by FirstNet to further expand its capabilities and usefulness.

Introduction

When a disaster occurs, first responders are dispatched to save lives, to protect property, and to assist in keeping order. To perform such duties, first responders require

L. S. Nagurney (✉)
Department of Electrical and Computer Engineering, University of Hartford, West Hartford, CT, USA
e-mail: nagurney@hartford.edu

© Springer Nature Switzerland AG 2018

I. S. Kotsireas et al. (eds.), *Dynamics of Disasters*, Springer Optimization and Its Applications 140, https://doi.org/10.1007/978-3-319-97442-2_7

operational communications among themselves, coordination communications with other agencies, and communications with the general population.

In this article, I describe the evolution of public safety disaster communications since the initial use of radio about 90 years ago to the current implementation of dedicated broadband spectra for first responders. I discuss some of the challenges and opportunities that face public safety communication as it has changed from traditional voice radio to broadband systems using smart phones, tablets, and advanced devices. In particular, I outline the development of the First Responder Network, FirstNet, in the US which is a prospective solution to many of the communications challenges, especially during disasters, for first responders. I then analyze the prognosis for FirstNet.

While wired communications, such as landline telephones, have been used since the latter half of the nineteenth century for disaster communications, the development of radio allowed untethered communications, thereby, removing geographical constraints on the responders. The use of radio, or wireless as it was then called, for police dispatch dates to the late 1920s when several radio enthusiasts in the Detroit Police Department installed a radio receiver in a police car. Such a one-way dispatch experiment continued for several years [8]. In 1933, the Bayonne, New Jersey police began to use two-way radios to dispatch officers [9]. In 1940, a major advance in communications came with the commissioning of a statewide two-way voice radio system for the Connecticut State Police [10]. Until the past decade or so most public safety radio communication continued to be verbal and most communication needs could be satisfied by analog radio systems. The major constraint for such systems on public safety, nevertheless, was the blocking of the radio channel due to too many messages being transferred at a given time.

Rudimentary digital data transmission systems for public safety were introduced in the late 1970s [5]. However, the physical layer of these systems continued to be based on analog radio systems. Most police digital queries of that time period consisted of text-based requests, involving the determination of the validity of a license, or the issuance of a warrant, and, typically, were received via a single line display in the police cruiser.

Over the years, the complexity and scope of incidents requiring first responder response have increased, with 9/11 being a prime example [42]. No longer is just a single police or fire department called to respond to a disaster, but, oftentimes, a multiplicity of agencies with common purposes are needed. In addition, the definition of first responder has been expanded from just local law enforcement and fire personnel to include those whose missions might include monitoring the environment, protection from biological and chemical hazards, and restoration of basic services. Each of these groups has specific communications requirements for their internal operations, as well as a need to communicate and share information with other responding agencies and stakeholders.

Another major change in communication resource needs for first responders is based upon the now universal use of smartphones, tablets, and other devices that connect to the internet. Many applications that first responders utilize, such as mapping, hazardous material identification and detection, local weather forecasts, etc., are not housed in a single device, or even a specific network, but are dependent on data available via the cloud. This sharing of data yields insights on the severity of the incident not only to the Incident Commander on the scene, but to all responders and their supervisors [24]. In the past decade the need for internet connectivity at an incident has moved from (a) not needed since we have proprietary systems that give us all the information, through (b) possibly useful, but not required, to (c) now, absolutely necessary for an incident to be efficiently managed.

Parallel to the growth and evolution of communication needs of first responders in a disaster are the communication demands of the news media and the communication needs of the humanitarian responders, who are involved in essential activities such as arranging for meals, shelter, and medical services for the displaced/victimized population. Furthermore, notifications of family and friends as to the welfare of survivors and possible reunification also require communications. While in earlier times such communications took place on separate radio channels and were delayed in time because of the limitations of existing technologies, currently, many, if not most, of these activities occur simultaneously to disaster response communications. Thus, with exploding demand from different stakeholders, there is a critical need for enhanced communication resources immediately post the disaster and as the recovery and reconstruction take place.

The communication bandwidth requirements post-disaster are schematically illustrated in Figure 1. As shown in the figure, the communications bandwidth requirements during a disaster increase rapidly during the first 24 hours, leveling and peaking during the first 3 days, and decreasing during the next 30+ days. This assumes, however, that it will be possible to restore much of the infrastructure during this period. For some emergencies and disasters this is, indeed, possible. However,

Fig. 1 Post-disaster communications bandwidth requirements

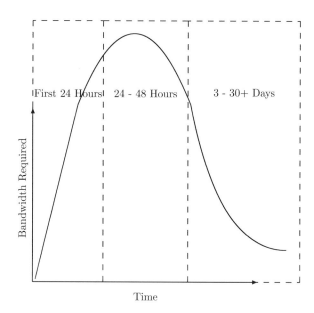

catastrophic disasters such as Hurricane Maria, which struck Puerto Rico and other Caribbean islands in September 2017, have become extended disasters where the realities of the first 3 days in terms of infrastructure losses following the disaster have been extended into the first 6 months [16].

It is important to note that wireless access to the internet for smart devices used by public safety personnel and first responders does not occur via dedicated public safety networks. Rather, it is provided by commercial wireless carriers such as Verizon, AT&T, T-Mobile, etc. Each carrier supplies broadband service based upon current broadband standards that can be accessed by devices whose owners have direct contracts with the cellular provider. Additionally, most carriers have roaming agreements that allow devices to shift to another carrier's network if there is no network availability from a device's contracted carrier.

Hence, the broadband wireless spectrum, shared among first responders, the media, humanitarian agencies, and the general public, is a scarce resource today with growing competition and increasing demands among users for the available bandwidth. For example, in the early tests of using commercial broadband for public safety at major events such as the Rose Parade in California on New Year's Day, it was noted that, when a parade float passed, many of the parade viewers would send their photos and videos of it; thereby, significantly degrading network performance [2]. Currently, there is a non-preemption status for any wireless user; consequently, a first responder needing to access building plans from the internet is competing with users sending low-priority messages. First responder communications are put in a queue with other users making it difficult for them to perform their functions efficiently and effectively.

Needs of First Responders

In 2015, Motorola surveyed public safety officials to determine their most pressing problems [29]. The major pressing problems were found to be:

- An increased level of community engagement needed;
- Real-time data should be accessible in the field;
- Increased communications with neighboring agencies are essential;
- Collaborative technologies must be used to expand capabilities;
- The technology skills of first responders must be better managed.

The solution to each of these pressing problems would likely require use of a broadband wireless service with broadband providing the high data rate needed and wireless allowing the users to be physically untethered from a network. For example, currently, police and fire departments can no longer just respond to calls and allow the news media to listen to their dispatch radios and view their call logs to determine which incidents are newsworthy and which of these rise to the importance of immediate reporting. Public safety agencies now routinely engage the community through the use of social media, including Twitter and Facebook, as the incident unfolds, to alert and inform the public and, in many cases, seek the public's help.

Moreover, as any incident unfolds the first responders need to access data. For example, in the case of a tank truck accident, the first responding unit on the scene can use broadband to instantaneously determine whether the incident is likely to escalate by determining the tank's contents through a search of the database of tank placards. An accident involving a truck that is leaking an inert, nontoxic gas, such as argon, has very low probability of impacting just beyond the accident site and, similarly, has a low probability of harm to first responders who take standard precautions. However, a truck leaking a chemical that generates toxic, explosive vapors must be managed significantly differently to prevent additional injury to the population and to the first responders themselves.

Keeping neighboring agencies informed of incidents and coordinating responses near their jurisdictional borders are other important tasks. Some incidents might require the immediate response from a neighboring agency based upon a mutual-aid agreement. Other incidents might be better managed through assistance in another jurisdiction. For example, a traffic jam due to a road closure near the incident could be minimized if motorists were encouraged in advance to take an alternate route. This could be very helpful in the case of evacuation networks (see, e.g., [46]).

Furthermore, the work of first responders has expanded to include many functions that were not even considered their responsibility only a few years ago. However, many of these functions are only infrequently performed and often require specialized skills and equipment. Collaboration, via the internet, can enhance every agency's capabilities and increase the availability of services and the efficient management of incidents.

The last pressing problem from the Motorola survey addresses the skills gap. The training requirements of first responders have grown exponentially. Recurrent training and certification on new techniques, equipment, and types of incidents have become mandatory. Enhancing and keeping current the appropriate skills for first responders are important concerns not only to their agencies, but to all.

Types of Disaster-Related Communications

The communications needs of first responders can, generally, be divided into two main categories: Mission Critical and Non-Mission Critical.

Mission Critical Communications are classified as those that have the highest urgency and need the maximum reliability [28]. Mission critical voice and data messages must be transmitted immediately with the lowest possible latency (delay). Currently, almost all mission critical communications among first responders are transmitted over channels dedicated to the primary agency. Police and fire dispatch are prime examples of mission critical communications. For situational awareness (cf. [24]), most mission critical communications are not single user to single user, but must be received simultaneously by all involved. Conventional push-to-talk radio, with each radio receiving all transmissions, is the most used technology for mission critical communications. The key metric that defines the performance of mission critical communications is reliability. Many first responders believe that simpler is better for mission critical communications.

Non-Mission Critical Communication includes an extensive class of messages that does not have to be transmitted with the highest urgency, but can be relegated to a lower priority. In a disaster, these non-life threatening or non-safety-of-life communications are important, but not so important that they must be transmitted with minimal latency. Messages, such as those involving logistics, slowly changing data, etc., are also included in this category. They are often from one individual to another without the broadcast requirement needed for situational awareness. While in many cases, non-mission critical communications are transmitted over the same communications channels as mission critical communications, they may also be transmitted over a secondary channel or by an entirely different mode such as cellphone, text message, or email. This is important in the case of the transmission of sensitive information, such as the names, personal information, and/or relevant medical information.

This paper is organized as follows. Section Brief Background on Communications Technology of this paper briefly reviews communications technology. Section Interoperability and Broadband discusses interoperability and broadband. Section History of FirstNet: First Responder Network Authority traces the history of the First Responders Network, FirstNet. Section Complementary Technologies discusses complementary technologies that should be used in conjunction with FirstNet. Section International First Responder Networks describes the advanced features needed in Long-Term Evolution (LTE) systems for first responders. Section Features Needed for LTE Networks to be More Functional for Mission Critical Communications presents the efforts in other countries for broadband for first responders. In the concluding Section Open Research Questions, open research questions regarding broadband for first responders are summarized.

Brief Background on Communications Technology

This section briefly reviews Land Mobile Radio (LMR) and LTE broadband networks.

Land Mobile Radio: LMR

Land Mobile Radio is also known as conventional 2-way radio communications that is ubiquitous in public safety agencies. The radios in police cruisers and fire trucks and the handheld radios that are strapped to the belts of our first responders are the prime examples of LMR. Almost all LMR systems are controlled directly by using agency but there is often no interoperability among systems. Within the US, most first responder dispatch uses legacy LMR systems. Despite new technology

Simplex LMR System   Repeater LMR System

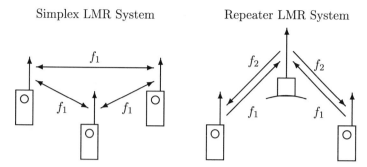

Fig. 2 Simplex and repeater LMR systems

and frequency bands becoming available, agencies often stick with LMR systems defined decades ago because they are mission critical and work. Many agencies do not replace their legacy LMR dispatch systems due to financial constraints. The legacy behavior limits interoperation, since, often, interoperation cannot occur by just changing the channel on a radio, but requires a whole different radio, antenna, and protocol [31, 33].

The simplest form of LMR, known as simplex, uses a single frequency, f_1, and is illustrated schematically on the left-hand side of Figure 2. Simplex radio's transmissions can be received by all other radios in the system simultaneously, similar to a party line phone. The two key limitations of simplex LMR are the power level of the radios, which often limits battery life, and the line-of-sight nature of radio propagation which introduces shadowing effects. Both of these couple to reduce the range of the transmissions. Simplex LMR systems, however, have the highest reliability, since no additional infrastructure, other than the radios and their associated antennas, is necessary for communications and user to user relay is possible. That is, if responder A cannot directly communicate with responder C, but responder B can communicate with both, responder B can serve as a relay between A and C [31, 33].

To extend the range of LMR systems and increase the spatial coverage, a repeater can be installed. A repeater is a receiver/transmitter pair. Users transmit to the repeater on one frequency, f_1. The repeater simultaneously retransmits (or repeats) the transmission on another frequency, f_2. The users' receivers are tuned to this frequency, f_2, and receive the transmission (see Figure 2). In this way, the limitations that can be attributed to lower transmitter power and shadowing are diminished. The repeater, which is often located at a remote location such as the top of a nearby mountain or tall building, must be operational for the system to work. Repeater sites almost always have redundant backup power systems so that the repeater continues operating even when primary electrical power is interrupted. At times, a hot-standby repeater is used to ensure service when the primary one becomes inoperable. Some systems include an additional standby repeater at another location to ensure highest reliability. Proper system design will often include an option for operation without the repeater for cases when the repeater is inoperative or units are out of range of the repeater.

One key limitation of conventional LMR is that the radio spectrum is a scarce resource. Conventional LMR requires a dedicated channel for each agency. Repeaterized systems require two channels per system. Radio channels may be reused with a spatial separation that is often measured in hundreds of miles. In addition, adjacent or even nearby channels may not be used in a close geographic area. One way to alleviate the spectrum shortage is *trunking*. A trunked system is based upon the premises that (1) each user agency normally uses only a relatively small amount of the channel capacity and (2) that the usage patterns of the users do not have significant overlap or are primarily just certain times of a day. Trunked systems allow multiple users to share a group of channels. While at times there might be a slight delay for a channel to become clear, the latency is not always noticed by the user. Although the trunking controller and repeater must be in operation for the system to work, with proper design, the system can continue to be operational if the controller/repeater is out of service.

While the association of LMR with a first responder's handheld radio implies that the system is analog, three digital protocols for LMR communication have been developed over the past two decades. In addition to transmitting voice, each can be configured to include short text messages (SMS) service and limited interoperability. Some even have the possibility of quasi-private user to user communication. These digital protocols include:

1. Terrestrial Trunked Radio (TETRA) is a European standard designed for public safety networks. For technical reasons, it does not meet the FCC-mandated spectral mask for use in public safety networks in the US, although several demonstration projects have been implemented. It is also extensively used for transportation networks in the US and worldwide and the military [13].
2. Project 25 (P25), also known as APCO-25, is a North American standard for digital radio communications for use by public safety organizations to enable interoperation. This standard is developed and promoted by the Association of Public-Safety Communications Officials, APCO. The standard is open so that any manufacturer can develop radios that conform to the standard [36]).
3. Digital Mobile Radio (DMR) is an open digital mobile radio standard defined by the European Telecommunications Standards Institute, ETSI, in 2005 and used internationally [11]. It was developed in tiers, the first for simple low-power, low-cost handheld radios in Europe. It has been extended to include talkgroups where groups of users can have partyline-like communications.

Long-Term Evolution: LTE

As cellular telephones became more popular and advancements were added so that users could send text messages, email, and access data services, the standards have evolved from the original analog cell standard, AMPS, through multiple generations. LTE was defined in the third generation standard to increase the data speed and to harmonize various data standards. Most wireless carriers now use LTE technology, including Verizon, which had previously used an incompatible standard. As in all cellular systems, the users' devices connect to cellsites (often known as towers). From these sites, the traffic is transmitted over the backhaul to the central office, known in LTE as the Evolved Packer Core (EPC). The EPC is an Internet Protocol (IP) based network that routes data among cellsites and connects devices to the internet. A key feature that led to the implementation of LTE networks is that each LTE cell can support approximately four times the data capacity when compared to previous generation of wireless networks [41].

Currently, LTE networks are clearly well-established in all markets. The needs of first responders require some extensions to LTE. These extensions are discussed in section Features Needed for LTE Networks to be More Functional for Mission Critical Communications.

LMR vs LTE

To summarize the differences between LMR and LTE, an LMR system uses preconfigured channels. Overlapping LMR coverage is provided by using different frequencies. The bandwidth and throughput of an LMR system are determined at the system design time. Users using their own LMR channel do not impact users of other LMR systems. In contrast, in the case of LTE, all sites operate on the same frequency; thus, the system must be designed and tuned to minimize overlapping coverage and the *channels* are dynamically managed at each site. The bandwidth and throughput are determined by need and availability to minimize congestion. One can envision an LTE system as one large data pipe containing individual data pipes. Near the site all pipes may be used together and the capacity is 74 Mbps. As a user moves away from the celltower, the capacity is reduced; however, the system can handle different communication demands.

Interoperability and Broadband

The 2015 Motorola Survey concluded that 78% of those surveyed desired easy interoperability with neighboring agencies and 73% desired the ability to connect different devices and networks together. This issue of interoperability was identified decades earlier by communications professionals. It was not until the well-publicized communications deficiencies during the World Trade Center attack on September 11, 2001, that interoperability issues became apparent to decision-makers and those responsible for allocating sufficient funding to mitigate the problem [42]. Almost simultaneous to the identification of the deficiencies of the traditional radio LMR systems, advanced devices such as smartphones and devices were thought to become a panacea. First responders embraced broadband using commercial technology and wireless networks. The growth and easy use of LTE networks for mobile broadband access became a natural for use by first responders. While LTE networks provide reliable, and now ubiquitous access, there are several drawbacks that prove difficult for first responder use.

Because the commercial carriers' business model is to satisfy as many users as possible, a goal of a wireless provider is to allow every user some access to the network. The primary drawbacks that this universal access causes include prioritization and data throttling. Despite the wireless providers' claims of unlimited data and equal priority in their advertising, the wireless bandwidth is still a limited resource. Most of the time, users do not realize this. If too many users are attempting to access the broadband network, some users' data transfer speed is throttled back. This choice is often made by the provider based upon a user's total data usage over a period of time. In addition, if a user uses above a certain threshold of data and the network is congested, the provider prioritizes other users whose usage has been less. To the first responder community, this is a major limitation, since, first, during an emergency incident, the network will become congested with users and the data needs of first responders will likely exceed the throttling limit.

In short, when a disaster occurs, broadband use (smartphone/tablet) increases to the point that the network becomes congested and throughput essentially goes to zero. First responders who have immediate need for the broadband network cannot access the network.

What Is Needed for Public Safety Broadband

After careful study, it was concluded that the broadband network needs of the first responders could be summarized as follows:

- Public safety broadband should use 700 MHz LTE to allow use of Commercial of the Shelf (COTS) devices;
- The network should be fully interoperable on a nationwide basis;
- Bandwidth will not be an issue for normal operation;
- While bandwidth becomes an issue when a large incident occurs, most incidents occur in a relatively confined geographically area involving only a small number of cell sectors. The network congestion is likely to occur in a limited region;
- Real-time network management will be required so that public safety will have preemptive priority.

First responders have tremendous communications challenges. Currently, in the US, there are over 10,000 radio networks dedicated to first responders (in the broadest sense). There are 3,100 counties and over 70,000 public service agencies in the US. In addition, there are over 550 recognized Native American tribes that perform their own public safety activities. A very rough estimate from these numbers indicates that there could be several hundred thousand devices used by first responders connected via the wireless network.

To satisfy these requirements, a national broadband network, operating in the 700 MHz LTE band was envisioned. The network could be built by a public–private partnership. While seed money from the government would be required, the network itself would be self-funding through network fees with the seed funding eventually returned to the government [14, 32].

History of FirstNet: First Responder Network Authority

To build, deploy, and operate a Nationwide Public Safety Broadband Network (NPSBN) based on a single, national network architecture, an independent federal authority with a statutory duty and responsibility to take all actions necessary was created as Section 6204 of the Middle Class Tax Relief and Job Creation Act of 2012, PL 112-96, approved February 22, 2012 [34]. The authority, named *First Responder Network Authority (FirstNet)*, was given the statutory responsibility to establish a national public safety broadband network, which includes not only the Core Network, but also the Radio Access Network (RAN) in each state or territory. For the purposes of the act there are 50 states and 6 territories, the District of Columbia, Puerto Rico, the US Virgin Islands, Guam, Micronesia, and Tribal Lands.

The statute also specifies that FirstNet create a Board of Directors with 15 members and establish advisory councils. It also requires that each governor of each state will appoint a Single Point of Contact to interface the state/territory with FirstNet. In addition, the statute created a technical implementation board to define the network. Band 14, a 20 MHz of spectrum dedicated nationwide for public safety in the 700 MHz LTE frequency range, was specified for the network with the law requiring the issuance of one nationwide license to FirstNet. The statute also specifies a 25-year life for FirstNet. The final network specifications included the use of 3GPP Standard Band 14 LTE with a 10 MHz wide uplink and a 10 MHz wide downlink. These frequencies had already been allocated for public safety use on a nationwide basis, but no licenses had been issued. To achieve higher reliability, a higher transmitter output power would be allowed for these Band14 devices, than is allowed for conventional smart devices.

The network will consist of the Radio Access Network (RAN), which is the collection of cellsites that the users access and the Extended Packet Core (EPC), the back office of the network. The system architecture of FirstNet is illustrated in Figure 3. While both will be built by FirstNet, the possibility that states could opt-out of the FirstNet RAN and build their own was considered and will be discussed in section Opt-Out Provision of FirstNet.

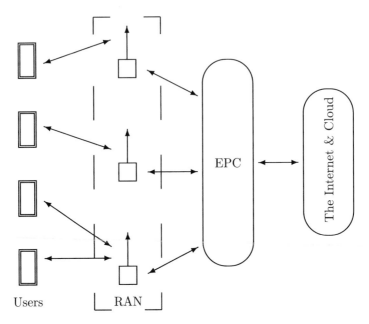

Fig. 3 FirstNet system architecture

Public–Private Partnership

The actual design, construction, and operational management would be delegated to a public–private partnership. The seed money, raised from spectrum auctions, would be used for the initial construction of the network. User fees on devices utilized by first responders and/or public safety on the network would pay for the operational expenses. In addition, during non-disaster situations, the partnership would be allowed to re-sell the network capacity to other users. At a value of $46.5 Billion, FirstNet is one of the largest public–private partnerships [38].

The contract was awarded on March 31, 2017 to an AT& T-led team that includes Motorola, General Dynamics, Inmarsat, and Sapient. Each of the partners has specific expertise it will bring to the project. Because AT& T currently manages a nationwide LTE system, the selection team concluded that they possess the necessary organizational and managerial skills for FirstNet to succeed. FirstNet will provide the 20 MHz of Band 14 spectrum and $6.5 billion in initial funding to the partnership. For a 25-year term, the AT&T-led consortium will deploy and operate the NSBN. During this time, AT&T is expected to spend about $40 billion for construction, operation, and maintenance of the network. The consortium will also ensure that the network evolves with the needs of first responders and with advances in technology.

When FirstNet's spectrum is not being used by public safety, AT&T may re-sell the network for other commercial purposes. First responders will be prioritized over any other commercial users on the Network. The contract will be overseen by the FirstNet Network Authority to ensure that the partners deliver innovation, appropriate technology, and customer care [17].

Opt-Out Provision of FirstNet

Probably the most controversial part of FirstNet was the opt-out provision. While the core network will be built by the FirstNet Partnership, states and territories would have the option to build their own Radio Access Networks (RAN), i.e., the cellsites and the associated backhaul. States who would opt-out of FirstNet would be required to construct their own Radio Access Network that would interface with the FirstNet core network and provide at least as good coverage as the consortium-built RAN would. As part of the FirstNet proof of concept, there were several demonstration networks, both temporary and quasipermanent, constructed to demonstrate FirstNet and validate its features. These would also be integrated into the final network.

The opt-out decisions quickly became contentious, since many states, at first, were not convinced that FirstNet was providing them enough information to make an informed decision. In addition, due to the differing of freedom-of-information and state government contract laws and regulations, some states did not believe that they could properly evaluate FirstNet's state plan for that state. While the statute prescribed funding for RAN construction and provisions for those states to license and control the FirstNet spectrum in their states, many states did not believe that the funding and restrictions were transparently outlined to them. Separately, it was also felt that the AT&T led partnership was playing hardball with states considering optingout. For example, AT&T offered states opting-in the ability to purchase devices that would connect with the current AT&T LTE network. AT&T would modify its current network to allow these public safety devices to preempt non-public safety users during an emergency [7, 48].

While a number of states considered opting-out, often hiring consultants to develop specifications for the alternate RAN that would be constructed at no cost to the state, by December 28, 2017, the final deadline to opt-out, all 50 states, the District of Columbia, Puerto Rico, the Virgin Islands, and several Pacific Territories opted-in [19]. New Hampshire was the only state that formally opted-out of the FirstNet RAN. However, when it became apparent that it would be the only state/territory that would opt-out, New Hampshire reversed its course and opted-in to a FirstNet constructed RAN [18, 44].

FirstNet Public Safety Users Quality of Service and Preemption (QPP)

In implementing FirstNet, the AT&T-led consortium divided public safety users into two tiers, primary and primary extended. Each tier will have different priority and preemption status. The monthly user fee for a device will depend on its tier; the primary extended tier device will have a lower monthly cost.

To speed up the implementation of FirstNet, because AT&T is a nationwide wireless carrier, the FirstNet consortium has immediately implemented first responder pre-emption on its current network. AT&T has indicated that FirstNet will be multiband and it will provide public safety with priority on all bands of its network, not just band 14 [1].

Primary users will include firefighters, police officers, and EMS. They will be the only network users who can actually preempt another user on the network. Primary public safety users will have a special access class. Users in this class will be exempt from throttling and barring. In addition, users in this class will have a high-priority access flag that will give them priority treatment during the various call setup stages. In times of high Band 14 traffic loads, non-primary public service users will be offloaded to other LTE bands [37].

Primary extended users will not have the preemption capabilities that primary users have; rather, they will still have priority status on the network.

Special incident management level priority will exist to allow any specific user to be lifted above other traffic for a specific period of time through a manual form of priority that users can provision through an incident management portal. Thus, should a primary extended user (or even just a normal user) be the first unit on the scene of an incident, that unit can assume the same properties as a primary user. It is expected that many agencies will have a much larger number of primary extended users because the price for these users will be less than the price for primary users.

Complementary Technologies

It is important to understand that FirstNet does not live in a bubble and there are complementary technologies that should be used to supplement first responders' needs. FirstNet is expected to last over two decades; thus, it is clear that the FirstNet of 2037 will probably look significantly different from the FirstNet of 2017.

ATSC 3.0

While FirstNet is designed for use by first responders, getting information to the affected population is also of major importance to first responders. For example, informing areas to evacuate and suggesting routes to take are important not only to the first responders, but also to the general public. Despite the social media efforts that first responders are currently ramping up, broadcast TV and radio can still play an important role in informing the public. Also, new TV standards include selective information transfer that can be used to complement FirstNet.

Our current broadcast emergency alert system in the US was developed during the Cold War to allow the President to address the country in times of emergency. It is based upon over-the-air broadcast technology which assumes that most of the population will have access to broadcast TV and radio. A key difference between broadcast and FirstNet type communications is that broadcast is a one-to-many communication without any response or acknowledgment. Even when IP type communication is used, there is some acknowledgment that the message is received, even if not acted upon. Because of the lack of acknowledgment or retransmission, broadcasting is inherently exempt from the congestion issues facing LTE networks. There is only one transmitting user on each channel who controls all material distributed. Under current broadcast technology, broadcasters can only send a single stream to all viewers, that is, all recipients of a broadcast station receive the same broadcast.

Broadcast transmissions are also highly reliable, since the transmitter site is designed for robustness and redundancy. All transmitter sites have an emergency source of power. There are usually multiple ways the studio can connect with the site. There is often a backup transmitter and antenna which allow the station to stay on the air throughout almost any disaster.

A next generation TV standard, ATSC 3.0, is currently being completed [4]. It expands the definition of television broadcasting to include not only traditional over-the-air television, but to include seamless integration of mobile devices. One key element of the standard is the ability to deliver multiple data streams that are tailored to specific segments. For example, multiple language support is one. The ability to combine an over-the-air stream to an internet distributed stream is another to achieve enhanced capability of the user experience is another. Each stream can use its own coding and modulation scheme therefore allowing almost unlimited options.

Part of this standard is an Advanced Emergency Alert (AEA) feature. This improves on the current emergency alert crawler, which is often seen on TV, but is often ignored since the alert is too general. ATSC 3.0 is designed to provide direct interaction between the TV broadcaster and the TV receiver. There will be a series of prompts and summaries on each device that will identify items of interest to that viewer. The user may select what alerts he/she will want. Rather than just the slow crawler at the bottom of the screen as currently done, there will be the ability to transmit graphics, multimedia, etc. Alerts can also be geotargeted to reach their primary audience in a given geographic area. Another feature is that devices can be woken up by an AEA alert, thus alerting those who do not even have their devices on. In a sense, this turns the device, TV, smartphone, tablet, etc., into the equivalent of an emergency pager.

It is expected that, when fully implemented, ATSC 3.0 will provide broadcasters with a more robust and reliable public warning and safety information communications system and leverage broadcaster's major role as public information provider during emergencies.

These multiple streams, the ability to target a stream to a small number of or even just a single user, and the ability to wake up devices will also allow the secure transmission of information and data to first responders. For example, a map or document might be sent over the broadcast channel to first responders without taking up the bandwidth of the LTE network. Because ATSC 3.0 works in the broadcast mode, users may modify their streams at any time [40].

It is important to realize that ATSC 3.0 technology is complementary and can be used collaboratively. Broadcasted data and video traffic, offloaded from the LTE system, during times of emergency will enable the LTE to do what it does best, mission critical communications [3].

Dedicated First Responder Wi-Fi

Despite all the promises for bandwidth availability from FirstNet, it is anticipated that first responders demands for bandwidth will soon exceed the supply. There are some applications whose communications needs can be better satisfied by other methods than an LTE network. For example, video surveillance is very bandwidth intensive and generates multimegabit-per-second data streams. The backhaul from fixed cameras can easily be fulfilled by a licensed microwave Wi-Fi system operating in the 4.9 GHz band licensed for public safety [30]. It will provide a high capacity network with low latency, ideal for such operations. Properly designed Wi-Fi nodes can become a redundant mesh network [26].

For temporary surveillance and extending radio coverage, a tethered drone could be used. Again, the 4.9 GHz band could be used to both to convey video from the drone and link it to the ground-based mesh network. This is important when the incident is out of range of many of the communications resources. Having both a radio range extender and video will enhance the capabilities of first responders [20].

T-Band and 700 MHz LMR

Despite all the communications advantages and new opportunities that FirstNet provides first responders, there still will be a need for mission critical voice communications separate from FirstNet for the forseeable future. In the creation of FirstNet, the statute required that the T-band, the radio spectrum between 470 and 512 MHz, be auctioned off, most likely for TV stations. The T-band is used for public safety communications in eleven major metropolitan areas and was originally used by TV channels 16–20. As spectrum congestion increased in other frequency bands in these areas, the T-band was allocated for public safety. However, by 2023, all users of the T-band will be required to relinquish their use of the band. The funds to replace the current T-band systems will be generated by the T-band spectrum auction so that there would be minimal financial impact on the current users or the government with this reallocation [15].

Clearly, the same pressures on the radio spectrum that led to the creation of the T-band are still faced by public safety agencies using T-band. The 700 MHz spectrum allocation for FirstNet includes some limited contiguous spectrum for LMR use by first responders. However, these frequencies cannot be used as direct replacements for T-band frequencies since equipment, propagation coverage, location of repeater sites, etc., are required to be different. Thus, the 700 MHz frequencies are not a direct replacement [43].

In the distant future, as Mission Critical Push-to-talk Voice-over-LTE and other extensions are created to enhance the LTE standard, it might be possible to reduce the reliance on LMR technology as it presently exists. However, until this happens, public safety agencies will continue to use LMR for much of their mission critical operations.

International First Responder Networks

Broadband for first responders is not just an issue in the US, but it is a worldwide issue. The problem becomes more pressing in many less developed countries when they host a major international event such as the Olympics. For an overall view of LTE networks for public safety worldwide, see [27] and [23].

Brazil, as the host for the 2014 World Cup and 2016 Olympics, provided an ideal demonstration site for LTE technologies and public safety. The Brazilian Army extended its LTE trial in 2014 in Rio de Janerio [47]. Additional demonstration LTE networks in major Brazilian cities were described in [45]. The Brazilian system was the first major event with broadband LTE. About 70% of the users were 4G [35].

England was one of the first countries to achieve interoperable voice communications between all emergency service users through a single system, Airwave. Developed in 2000, the Airwave network is a public–private partnership. Currently, England is now the first country to fully replace its singular nationwide public safety network [23]). It is expected that the network will be operational by 2020 [21].

South Korea, the host of the 2018 Winter Olympics will begin nationwide deployment of its public-safety Long-Term Evolution (LTE) network with projects including one for the Olympics. The country will be investing about US$880 million to deploy the network and US$900 million for 10 years of operation following pilot projects in 2015 and 2016. Much of the nationwide network will be completed by 2020 [49].

Canada, because of its long common border and need for interoperability with the US will build a Band 14 LTE network that is similar to FirstNet. One question that has been raised in Canada is that, due to security concerns, equipment by several vendors has been forbidden to be used on the US FirstNet. This raises some questions on how interoperability will occur [6].

Other countries including Australia, Hong Kong, China, Chile, Finland, and France are in the planning, pilot, or early operational stage of first responder broadband.

Features Needed for LTE Networks to be More Functional for Mission Critical Communications

Although an LTE-based FirstNet will be a boom for public safety, there are several enhancements or extensions to the LTE standard that would make an LTE network even more useful and that would remove some of the limitations of the LTE network. These are needed especially when mission critical communications will be transmitted over the LTE network and reliability will be paramount.

The first needed enhancement is *Proximity Services*. Smart devices using LTE require a cellsite to communicate with each other even if the devices are in close proximity. Without an active cellsite two LTE devices cannot communicate with each other. Proximity services will allow devices on the network to identify other devices in close physical proximity and enable them to make device-to-device communications even when the LTE network is not functional or both devices are out of range of a cellsite [2].

Because first responders might use their devices in locations that are not in line of sight with the cell towers, a second enhancement, *User Equipment to Network Relay*, must be implemented. This feature will allow a mobile device to serve as a relay for other devices to the nearest cellsite and, thus, provide connectivity for these out of range devices. In addition, this feature allows mobiles to act as relay points between other mobiles, permitting communication without going via the network even if the mobiles are out of direct range [2].

The full convergence of a public safety broadband network and LMR will not occur until *Mission Critical Push-To-Talk (MCPTT)* is specified and standardized. MCPTT will allow mission critical push-to-talk LMR style communications over an LTE network. The standardization is important since most public safety users do not want to be constrained to one vendor, but would like to be able to mix and match vendors of the PTT devices. This is analogous to current LMR practice, where, in most cases, use of open standards allows purchase of radios from many vendors. In June 2017 an interoperability test of using equipment from over 20 vendors achieved an 85% success rate after over 1000 tests [12].

FirstNet will also require enhanced security to protect the system from unauthorized users, eavesdropping, denial of service attacks, and other security risks.

Open Research Questions

While a base FirstNet will be implemented during 2018, there are still a variety of research questions that need to be answered as the system evolves over the next quarter century. While many research questions involve improvement of the LTE standard itself as discussed in the previous section, there are others that do not directly relate to the LTE standard [25, 39]. These include:

Cellular Systems as Self-organizing Networks Despite the care taken in designing a network, it is likely that, at some point, one or more cellsites will be off the air and unusable. When this happens in a normal cellular network, often the fringe coverage for adjacent cellsites is sufficient that, while the coverage is degraded, it is still acceptable in an outdoor environment. In an urban environment where first responders are expected to be inside a building, this loss of performance will be noticeable. In rural areas there will be a much higher likelihood of service disruption when a cellsite is inoperable.

Understanding Public Safety Grade While cellsites themselves and the network, in general, are built to 99.999% reliability, the outages in disasters are local and could affect a large population. For example, Ross et al. [39] notes that after Hurricane Irma 50% of the sites in Miami-Dade County were knocked out. While these disruptions would have had little effect on the performance metrics of FirstNet, they had a significant effect on a large population.

Post LTE Roadmap As broadband network standards evolve, so must FirstNet. The next generation of broadband standards, 5G, are being finalized. How will FirstNet evolve to use these standards? During AT&T's Quarterly Earnings call in January 2018, its Chairman and CEO Randall Stephenson said: "What we're doing is building a nationwide network with the latest technology. It's designed and it's hardened for America's first responders, and then that will be our foundation for broad 5G deployment" [22].

The final question is *Is the political/business model sustainable?* Although public–private partnerships have been around for a long time, the size and scope of this project may challenge these.

Acknowledgements I would like to thank the organizers of the 3rd Dynamics of Disasters Conference in Kalamata, Greece in July 2017 for organizing such a stimulating conference in an outstanding venue. I would also like to thank Madeleine Noland and John Lawson for helpful discussions on ATSC 3.0. Finally, the constructive comments of an anonymous reviewer and the Editors on an earlier version of this paper are appreciated.

References

1. AT&T: FirstNet Launches Ruthless Preemption for First Responders, News Release, 12 Dec 2017. http://about.att.com/story/preemption_for_first_responders.html
2. Bishop, D.: The evolution of public safety: FirstNet and LTE. Above Ground Level **14**(9), 18–20 (2017)
3. Chernock, R.: Advanced emergency alerting, NextGenTV Conference, 16–17 May 2017. https://www.atsc.org/pdf/conference/Chernock_AEA.pdf.
4. Chernock, R., Whitaker, J.C., Wu, Y.: ATSC 3.0 the next step in the evolution of digital television. IEEE Trans. Braodcasting **63**(1), 166–169 (2017)
5. Corillo, G.F.: The Effect of the Mobile Data Terminal on the Virginia Beach Police Department and the Impact of Technology on the History of Law Enforcement, Ph.D. Thesis, Regent University, Virginia Beach (2003)
6. Defence Research and Development Canada (DRDC): Implementation Models for a Public Safety Broadband Network, Scientific Letter DRDC-RDDC-2017-L121, 19 Apr 2017. http://cradpdf.drdc-rddc.gc.ca/PDFS/unc276/p805271_A1b.pdf

7. Engebretson, J.: FirstNet Opt Out: With Accusations Flying, AT&T and FirstNet Respond. Telecompetitor.com posted 30 Oct 2017. http://www.telecompetitor.com/firstnet-opt-out-with-accusations-flying-att-and-firstnet-respond/

8. ETHW: Milestones:One-Way Police Radio Communication, 1928, IEEE Southeastern Michigan Section Engineering Milestone Plaque, Detroit, MI, dedicated May 1987, ETHW wiki, 31 Dec 2015. http://ethw.org/Milestones:One-Way_Police_Radio_Communication,_1928.

9. ETHW: Milestones: Two-Way Police Radio Communications 1933, New Jersey Section IEEE, Engineering Milestone Plaque, Bayonne, MI, dedicated May 1987, ETHW Wiki, 31 Dec 2015. http://ethw.org/Milestones:Two-Way_Police_Radio_Communication,_1933

10. ETHW: Milestones: FM Police Radio Communication, 1940, Connecticut Section IEEE, Engineering Milestone Plaque, Hartford, CT, dedicated June 1987, ETHW Wiki, 17 Apr 2017. http://ethw.org/Milestones:FM_Police_Radio_Communication,_1940.

11. ETSI: Electromagnetic Compatibility and Radio Spectrum Matters (ERM); Digital Mobile Radio (DMR) General System Design ETSI TR 102 398 V1.3.1 (2013-01) (2013). http://www.etsi.org/deliver/etsi_tr/102300_102399/102398/01.03.01_60/tr_102398v010301p.pdf

12. ETSI: First ETSI LTE Mission-Critical Push to Talk Interoperability Tests Achieve 85% Success Rate, ETSI News Release, 26 June 2017. http://www.etsi.org/news-events/news/1201-2017-06-news-first-etsi-lte-mission-critical-push-to-talk-interoperability-tests-achieve-85-success-rate

13. ETSI: TETRA, http://www.etsi.org/technologies-clusters/technologies/tetra. Accessed 3 Feb 2018.

14. Farrill, F.C.: FirstNet Nationwide Network (FNN) Proposal, First Responders Network Authority Presentation to the Board, 25 Sep 2012. https://www.ntia.doc.gov/files/ntia/publications/firstnet_fnn_presentation_09-25-2012_final.pdf

15. Federal Communications Commission: Public Safety T-Band Fact Sheet July 2016. https://transition.fcc.gov/pshs/docs/T-Band_FactSheet_July2016.pdf

16. Federal Emergency Management Agency: Hurricane Maria Update, Release: DR-4339-PR NR 046, 6 Nov 2017. https://www.fema.gov/news-release/2017/11/06/4339/hurricane-maria-update

17. FirstNet: FirstNet: Top 10 Frequently Asked Questions (2017). https://www.firstnet.gov/sites/default/files/FirstNet_Partnership_FAQs__0.pdf

18. FirstNet: New Hampshire to Transform Communications for Public Safety; Governor Sununu Approves Buildout Plan for First Responder Network, Press Release, 28 Dec 2017. https://firstnet.gov/news/new-hampshire-transform-communications-public-safety

19. FirstNet: First Responder Network Goes Nationwide As All 50 States, 2 Territories and District of Columbia Join FirstNet, Press Release, 29 Dec 2017. https://firstnet.gov/news/first-responder-network-goes-nationwide

20. Gilbreth, D.: Use an instant tower for aerial policing, a new trend to supplement public safety. Above Ground Level 14(11), 10–13 (2017)

21. Jackson, D.: UK Public-safety LTE Network Targeted for Completion in 2020, with LMR Providing Early Direct-mode Solution, Urgent Communications. Posted online 7 July 2017. http://urgentcomm.com/public-safety-broadbandfirstnet/uk-public-safety-lte-network-targeted-completion-2020-lmr-providing-

22. Jackson, D.: AT&T CEO Highlights FirstNet Role as Foundation to Carrier's 5G Plans, Urgent Communications. Posted online. 1 Feb 2018. http://urgentcomm.com/public-safety-broadbandfirstnet/att-ceo-highlights-firstnet-role-foundation-carrier-s-5g-plans

23. Kable Business Intelligence: First Responder Solutions in the UK and Internationally, Report to the National (UK) Audit Office (NAO) Emergency Services Mobile Communications Programme (ESMCP). 22 Sep 2016. https://www.nao.org.uk/wp-content/uploads/2016/09/First-Responder-Solutions-in-the-UK-and-Internationally.pdf.

24. Karagiannis, G.M., Synolakis, C.E.: Collabrative incident planning and the common operational picture. In: Kotsireas, I.S., Nagurney, A., Pardalos, P.M. (eds.) Dynamics of Disasters – Key Concepts, Models, Algorithms, and Insights, pp 91–112. Springer International Publishing, Switzerland (2016)

25. Kindelspire, C.: The LMR to LTE transition. Mission Critical Commun. **32**(10), 18–21 (2017)
26. Lambert, L.: Public safety broadband communications is more than LTE. Above Ground Level **14**(11), 6–8 (2017)
27. Lynch, T.: LTE in Public Safety, IHS Technology Whitepaper, April 2014. https://technology.ihs.com/api/binary/580535?attachment=true
28. Motorola: Mission Critical Communications Designed to a Tougher Standard, Motorola Solutions White Paper 2013. https://www.motorolasolutions.com/content/dam/msi/docs/en-xw/static_files/Mission_Critical_Communications_Tougher_White_Paper.pdf
29. Motorola: 5 Trends Transforming Public Safety Communications: 2015 Nationwide Public Safety Industry Study Results. Motorola Solutions White Paper, 11-2015. https://www.motorolasolutions.com/content/dam/msi/docs/2015_public_safety_survey_white_paper.pdf
30. Motorola Spectrum Strategy: 4.9 GHz Public Safety Broadband Spectrum Overview of Technical Rules And Step-By-Step Licensing Instructions Motorola, Inc. July 26 2010. https://www.apcointl.org/spectrum-management/resources/spectrum/2011-12-08-22-11-29/file.html
31. Noll, E.M.: Landmobile and Marine Mobile Technical Handbook. Howard W. Sams, Indianapolis (1985)
32. NTIA: National Telecommunications and Information Administration Notice of inquiry: Development of the Nationwide Interoperable Public Safety Broadband Network. Federal Register **77**(193), 6080-6081, 4 Oct 2012. https://www.ntia.doc.gov/files/ntia/publications/firstnet_noi_10042012.pdf
33. Orr, W.I.: Radio Handbook, 27th ed. Howard W. Sams, Indianapolis (1981)
34. PL 112-96: Middle Class Tax Relief and Job Creation Act of 2012, PUBLIC LAW 11296. https://www.gpo.gov/fdsys/pkg/PLAW-112publ96/pdf/PLAW-112publ96.pdf
35. Prescott, R.: Rio Olympics: Mobile Carriers Prep for Heavy Social Media Use, RCR Wireless News, posted online 20 July 2016. https://www.rcrwireless.com/20160720/americas/rio-olympics-mobile-carriers-tag5
36. Project25 Interest Group: http://www.project25.org/
37. Ramey, D.: AT&T, FirstNet officials detail public-safety LTE priority, pre-emption capabilities. Mission Critical Commun. Published Online 14 June 2017. https://www.rrmediagroup.com/Features/FeaturesDetails/FID/758/
38. Ross, W.: U.S. Secretary of Commerce Wilbur Ross Announces FirstNet Public-Private Partnership, 30 Mar 2017. https://www.commerce.gov/news/secretary-speeches/2017/03/us-secretary-commerce-wilbur-ross-announces-firstnet-public-private
39. Ross, J., Sidore, S., Edison, S., Pao, T.: The challenge of public safety grade LTE. Mission Critical Commun. **32**(9), 16–21 (2017)
40. Siegler, D.: Changing Channels: How ATSC 3.0 Revolutionizes Broadcast TV. IEEE Broadcast Technology, Fourth Quarter, 51–53 (2017)
41. Signals Research Group: The LTE Standard, April 2014. https://www.qualcomm.com/media/documents/files/the-lte-standard.pdf
42. Simon, R., Teperman, S.: The World Trade Center Attack: lessons for disaster management Critical Care **5**(6), 318–320 (2001). Published online 6 Nov 2001. https://doi.org/10.1186/cc1060. https://www.ncbi.nlm.nih.gov/pmc/articles/PMC137379/
43. Springer, B.: One State's Perspective on T-Band, Mission Critical Communications, November-December 2017, 8
44. Sununu, C.: Governor Sununu Announces Decision to Opt Out of FirstNet Plan, Pursue Innovative Alternative Plan, Press Release, 7 Dec 2017. https://www.governor.nh.gov/news-media/press-2017/20171207-opt-out.htm
45. Tetra Today: Motorola Solutions Demonstrates LTE for Mission-Critical Comms in Brazil, posted 20 July 2016. http://www.tetratoday.com/news/motorola-solutions-demonstrates-lte-for-missioncritical-comms-in-brazil
46. Vogiatzis, C., Pardalos, P.M.: Evacuation modeling and betweenness centrality. In: Kotsireas, I.S., Nagurney, A., Pardalos, P.M. (eds.) Dynamics of Disasters – Key Concepts, Models, Algorithms, and Insights, pp 345–360. Springer International Publishing, Switzerland (2016)

47. Wendelken, S.: Wireless Plays Vital Role for World Cup Public Safety, Fan Connections. Mission Critical Commun. posted online 30 Jun 2014. https://www.rrmediagroup.com/Features/FeaturesDetails/FID/465
48. Wendelken, S.: FirstNet Rejects FOIA Request for State Plan Information, AT&T Contracts. Mission Critical Commun. Posted online 12 Sep 2017. https://www.rrmediagroup.com/Features/FeaturesDetails/FID/788
49. Wendelken, S.: South Korea to Launch Initial Public-Safety LTE Services in 2018. Mission Critical Commun. Posted online 16 Oct 2017. https://www.rrmediagroup.com/Features/FeaturesDetails/FID/794

A Humanitarian Logistics Case Study for the Intermediary Phase Accommodation Center for Refugees and Other Humanitarian Disaster Victims

Sofia Papadaki, Georgios Banias, Charisios Achillas, Dimitris Aidonis, Dimitris Folinas, Dionysis Bochtis, and Stamatis Papangelou

Abstract The growing and uncontrollable stream of refugees from Middle East and North Africa has created considerable pressure to governments and societies all over Europe. To establish the theoretical framework, the concept of humanitarian logistics is briefly examined in this paper. Historical data from the nineteenth century onwards illuminates the fact that this influx is not a novelty in the European continent and the interpretation of statistical data highlights the characteristics and particularities of the current refugee wave, as well as the possible repercussions these could inflict both to hosting societies and to displaced populations. Finally, a

S. Papadaki (✉)
International Hellenic University, School of Economics, Business Administration and Legal Studies, Thermi, Greece
e-mail: s.papadaki@ihu.edu.gr

G. Banias
International Hellenic University, School of Economics, Business Administration and Legal Studies, Thermi, Greece

Centre for Research and Technology – Hellas, Institute for Bio-economy and Agri-technology, Thermi, Thessaloniki, Greece

Ch. Achillas
International Hellenic University, School of Economics, Business Administration and Legal Studies, Thermi, Greece

Department of Logistics, Technological Educational Institute of Central Macedonia, Branch of Katerini, Katerini, Greece

D. Aidonis · D. Folinas
Department of Logistics, Technological Educational Institute of Central Macedonia, Branch of Katerini, Katerini, Greece

D. Bochtis
Centre for Research and Technology – Hellas, Institute for Bio-economy and Agri-technology, Thermi, Thessaloniki, Greece

S. Papangelou
University of Macedonia, School of Economics, Thessaloniki, Greece

© Springer Nature Switzerland AG 2018
I. S. Kotsireas et al. (eds.), *Dynamics of Disasters*, Springer Optimization and Its Applications 140, https://doi.org/10.1007/978-3-319-97442-2_8

review of European and national legislation and policies shows that measures taken so far are disjointed and that no complete but at the same time fair and humanitarian management strategy exists.

Within this context, the paper elaborates on the development of a compact accommodation center made of shipping containers, to function as one of the initial stages in adaptation before full social integration of the displaced populations. It aims at maximizing the respect for human rights and values while minimizing the impact on society and on the environment. Some of the humanitarian and ecological issues discussed are: integration of medical, educational, religious and social functions within the unit, optimal land utilization, renewable energy use, and waste management infrastructures. Creating added value for the "raw" material (shipping containers) and prolonging the unit's life span by enabling transformation and change of use, transportation and reuse, and finally end-of-life dismantlement and recycling also lie within the scope of the project.

The overall goal is not only to address the current needs stemming from the refugee crisis, but also to develop a project versatile enough to be adapted for implementation on further social groups in need of support. The paper's results could serve as a useful tool for governments and organizations to better plan ahead and respond fast and efficiently not only in regard to the present humanitarian emergency, but also in any possible similar major disaster situation, including the potential consequences of climate change.

Introduction

Today, the total number of forcibly displaced persons globally has reached over 65 million (Figure 1). In situations like these, the most important thing in every step of the way is the existence of basic support structures within the frame of humanitarian logistics [2].

The refugee crisis experienced today in Europe is by no means a novelty. The nineteenth and twentieth century have seen a number of significant migration waves,

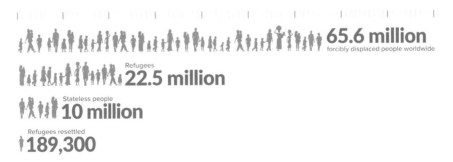

Fig. 1 Current numbers of forcibly displaced people, refugees, and stateless people worldwide [1]

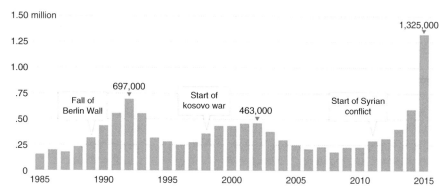

Fig. 2 Number of asylum seekers in Europe from 1985 until 2015 [6]

both from and towards Europe [3, 4]. The fact that discriminates the twenty-first century from the past is the simultaneous existence of several conflict areas. In the European "neighbourhood" alone, there are armed struggles in the Middle East, in North Africa, and in some parts of sub-Saharan Africa as well [5]. These conflicts have caused the vulnerable populations to flee their countries and seek a better future in the supposedly more stable Europe. According to the predictions, war and destruction of all infrastructure in those parts of the world will not come to an end anytime soon. Only during 2015, over one million people came to the European Union to find safety, increasing the total number of refugees in this region to almost 4.4 million (Figure 2). The majority of them is originating from Syria, Afghanistan, and Eritrea [7, 8], almost one in five are adult women and at least one in four are children, many of whom unaccompanied [9]. They follow travel routes well established by human-trafficking groups and yet anything but safe and secure.

Even after reaching the perceived safety of Europe, refugees still have to face significant integration issues [4], while being at the same time the trigger to xenophobia and political extremism. Europe has only recently managed to slow down the inflow of refugees, due to increased border control and mostly sporadic, unilateral actions and unfortunately not because of a comprehensive common strategic framework actually resolving all aspects of this crisis [10]. One of the most significant issues stemming from this crisis, and the main focus of this paper, is the problem of shelter as an essential human need for the refugees.

Definition of Migration

Migration is and always has been a fact of life for all living organisms. Whenever the living conditions become intolerable, a mass movement of populations occurs [11]. Human migration causes are varied and complex and include, among other things, economic or educational opportunities, wars or civil unrest, human rights abuse, as well as environmental reasons [12, 13]. Although quite often migration is caused simply by the quest for better work, yet in many cases it is a central part of complex humanitarian emergencies.

Included within the general framework term of migration and migrants, the refugees form their own distinct category. They are characterized as victims, dependent on others, and in need of humanitarian assistance [14]. These are people who were forced to flee their own country and seek international protection, due to human rights abuse and the inability of their governments to protect them [12]. Within this context, an asylum seeker is defined as a person who has left their country and has requested international protection but has not yet been granted refugee status [12, 14]. The definition coined by the International Organization of Migration [15] is considered at present the most successful in integrating all aspects of migration; according to it, a migrant is:"...any person who is moving or has moved across an international border or within a State away from his/her habitual place of residence, regardless of (1) the person's legal status; (2) whether the movement is voluntary or involuntary; (3) what the causes for the movement are; or (4) what the length of the stay is."

The Concept of Humanitarian Logistics

It was during the Indian Ocean tsunami relief operation in 2004 that logistics became for the first time the epicenter of attention as an integral part of any humanitarian relief operation [2, 16]. In fact, humanitarian relief operations are becoming increasingly demanded, since basic everyday goods, such as water and food, safety and shelter, health and education are in scarcity in many parts of the world. At the same time, natural and man-made disasters are occurring nowadays with alarming frequency [7]. As presented in Figure 3, disasters are divided into four distinct categories according to their cause and speed of occurrence [2]: (1) calamities: earthquakes, tornadoes, hurricanes, (2) destructive actions: industrial accidents, terrorist attacks, (3) plagues: poverty, famine, draught, and (4) crises: political or refugee crises.

Even though the relevance of logistics effort might vary depending on the type of disaster, its importance is by no means debated [2]. As a result, it has by now been established that at the core of an effective and efficient response to any humanitarian emergency lies an intertwined range of activities best known as humanitarian logistics. It becomes clear that this concept includes much more than simple material goods and their transport and distribution. Advance preparation, coordination of human resources, collection and processing of data, and extensive use of expert knowledge should all form an integral part of this process, in order to achieve— beyond logistic performance—a holistic supply chain management as well [2, 17]. In that respect, out of the four phases comprising a disaster management cycle— mitigation, preparation, response, and reconstruction—humanitarian logistics and supply chain management can and should be integrated in three, with mitigation the

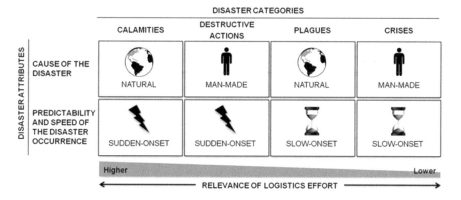

Fig. 3 Types of disasters [2]

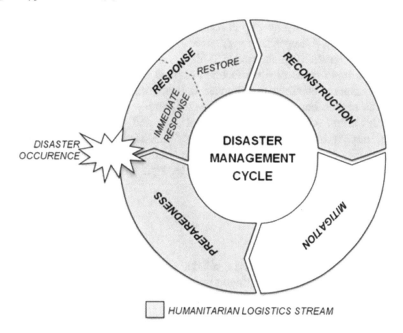

Fig. 4 Disaster management and the humanitarian logistics stream [2]

only exclusion. In fact, as depicted in Figure 4, the actual disaster occurrence is by no means the beginning of the cycle [2].

When addressing refugee crises similar to the one Europe is confronted with, humanitarian logistics can prove to be essential, both in the transition stage—providing the displaced people with the means for their safe and uninterrupted transport to safe areas—and in the initial accommodation stage—ensuring that the refugees' basic needs for food, water, sanitation, and shelter can be satisfied with adequacy and dignity [16, 18].

Unfortunately, the demand for humanitarian disaster relief will continue to grow. According to the predictions, for the next 50 years and due to political, social, economical, environmental, and health causes, disaster situations are expected to quintuple [16], thus making humanitarian logistics an absolute necessity.

Historical and Political Background of the Refugee Crisis in Europe

Push and Pull Factors

The images of hundreds of thousands of refugees desperately seeking shelter and being dismissed are not a novelty in recent history. Throughout the nineteenth and twentieth century, Europe has experienced six significant migration waves with diverse characteristics [3], outlined in Table 1.

The two main types of push factors observed in the last two decades of the twentieth century continue to influence the present migration waves and force people out of their home countries [19] are (a) extreme poverty in several countries of North-Central Africa and South-Central Asia and (b) the armed conflicts in Eritrea, Iraq, Afghanistan, and more recently Syria. Moreover, the perceived safety and prosperity of Europe acts as a very strong pull factor; especially countries, like Germany and Sweden, distinguished for their high levels of social services and support [20].

Future Predictions

Due to the complexity of this issue, reliable predictions regarding future refugee flows to Europe are extremely difficult to make [7]. In strictly numerical terms, the present rate of refugee arrivals is not unparalleled. However, refugee flows from previous occasions were either large but progressive, or rapid but moderate and therefore in both cases easily controllable. Such an extensive and at the same time sudden and unanticipated tide of war refugees is unprecedented. Moreover, there are several other factors that need to be taken into consideration.

The continuing war in Syria leaves no room for optimism. The projection for future refugee flows is theoretically ten times larger than the number of already departed persons. The situation in Afghanistan and Iraq is still anything but stable. In fact, the whole Middle East region has been historically suffering from territorial disputes that seem unable to be resolved in the near future, thus increasing the risk for new refugee flows [21]. The current refugee trend has the potential to influence further displaced populations so far uninterested to move away from their regions

Table 1 Main phases of migration in Europe in the nineteenth and twentieth century (based on data from [3])

Phase	1 Nineteenth century	2 1880s on	3 1914 on	4 1945–1950s	5 1960s–1970s	6(a) 1980s	6(b) 1990s
Source of refugees	• Europe	• Europe	• Europe	• Europe	• LDCs • East Europe	• LDCs	• Europe
Destination of refugees	• Europe	• Outside Europe	• Outside Europe • Europe	• Outside Europe • Europe	• Europe by invitation	• Europe	• Europe • outside Europe
Numbers average/annum (in thousands)	• Small	• 71	• 380	• 2500	• 13	• 70	• 600
Characteristics of refugees	• European • Exiles • Well-to-do	• Europeans • Jews • Impoverished	• Europeans • Jews • Impoverished	• Europeans • Jew/Russian • Cross-section	• Non-European • Impoverished	• Non-European • Impoverished	• Europeans • Impoverished • Skilled
Cause of flight	• Political	• Ethnic • Economic	• Genocide • Political	• Aftermath of war	• Decolonization • Ethnic • Political	• Political • War • Famine • Ethnic • economic	• Political • War • Ethnic
Organizational structure	• No	• Community based NGOs	• Ad hoc Eurocentric • International	• Eurocentric • International	• Ad hoc governmental • UN mediating	• Pan-European • NGOs	• Pan-European • NGOs
Government response	• No	• Legislation	• Intervention restriction	• Active recruitment	• Intervention restriction	• Exclusion	• Panic • Exclusion

into seeking asylum in Europe, as indicated by the recent flair in asylum claims from Palestinians [7].

Concept Development and Key Research Questions

Key Research Questions

The present crisis is considered to be more complex than others, because it incorporates humanitarian, political, religious, financial, and social parameters in both source and host countries alike. After analyzing and evaluating the profile of the situation, one of the most significant issues to be identified is the problem of shelter as an essential human need for the refugees. The paper elaborates on the development of a compact accommodation center made of shipping containers, to function as one of the initial stages in adaptation before full social integration of the displaced populations. It aims at maximizing the respect for human rights and values while minimizing the impact on society and on the environment. Some of the humanitarian and ecological issues discussed are: integration of medical, educational, religious and social functions within the unit, optimal land utilization, renewable energy use, and waste management infrastructures. Creating added value for the "raw" material (shipping containers) and prolonging the unit's life span by enabling transformation and change of use, transportation and reuse, and finally end-of-life dismantlement and recycling also lie within the scope of the project. Within this context, the key research questions that this paper attempts to negotiate with are the following:

(a) Examine the common strategies and practice adopted in similar circumstances on a global scale in general and in Europe in particular.
(b) Determine the key issues that need to be incorporated in any housing development attempt.
(c) Discuss the possibility of improving the current situation by featuring these key prerequisites in a more targeted approach to accommodation.

The basic aim of this paper is the creation of a module that can address the current situation and at the same time be adaptable to any further future humanitarian disaster, thus contributing to an essential level of preparedness.

Research Methodology

In order to address these research topics, an extensive literature review has been conducted; more specifically, a critical analysis of the available literature that has been found in Elsevier, Science Direct, Springer, The Lancet, as well as reports from EU Agencies (including Frontex and Europol), US Agencies, and organizations such as UNHCR, International Organization of Migration, Amnesty International,

Médecins Sans Frontières, Asylum in Europe, Forced Migration Review, International Institute for Strategic Studies, Center for Strategic and International Studies, PEW Research Center, and International Monetary Fund, focusing in the following areas of interest:

1. Scientific work regarding the humanitarian logistics of refugee crises.
2. Demographic and numerical data related to the refugee populations.
3. Generally accepted guidelines and minimum requirements for shelter design and construction, established by a variety of sources.

Concept Development

The literature review revealed a significant gap, both in theory and in practice, with respect to a rational, multi-functional, and inclusive housing solution for displaced persons. This finding appealed particularly to the authors' scientific and cognitive background in architecture and engineering. As a result, this paper focuses on the idea of an exemplary accommodation center, equipped with all necessary services to provide refugees with basic coping skills and aiming to act as an intermediary step between first reception until full social integration. It is created as a flexible structure following the existing guidelines and design criteria for such settlements while furthermore incorporating sustainable tactics regarding energy consumption and waste management.

Legislative Framework, Practices, and Humanitarian Logistics

Existing Numerical and Statistical Data for Europe and the Middle East and North Africa Region

In 2015 the number of forcibly displaced people globally surpassed 60 million, a number exceeding the population of the United Kingdom and reaching almost 1% of the global population; as a comparison, a country comprised of all these people would be ranked 21st largest in the world (Figure 5). Compared to the 42.5 million in 2011, this signifies an escalation of over 50% in just 5 years [8].

Figure 6 illustrates, for the year 2015, a significant advancement in total refugee populations worldwide; especially in Europe, over 1.3 million new refugees were recorded, a number translating in a radical increase of 43% [8].

The Middle East and North Africa (MENA) regions are the two geographical areas mainly accountable for the influx of refugees in Europe during the last years [23]. However, the impressive fact, as seen in Figure 7, is that more than half of the global refugee flows originate in just three countries of this region [1].

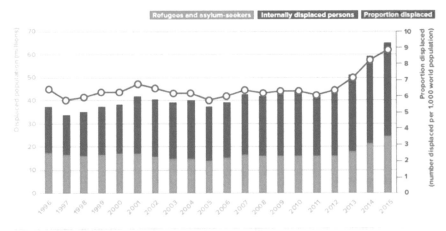

Fig. 5 Trend of global displacement and proportion displaced between 1996 and 2015 (end of year) [22]

UNHCR regions	Start-2015			End-2015			Change (total)	
	Refugees	People in refugee-like situations	Total refugees	Refugees	People in refugee-like situations	Total refugees	Absolute	%
- Central Africa and Great Lakes	625,000	37,600	662,600	1,173,400	15,900	1,189,300	526,700	79
- East and Horn of Africa	2,568,000	33,400	2,601,400	2,739,400	-	2,739,400	138,000	5
- Southern Africa	177,700	-	177,700	189,800	-	189,800	12,100	7
- West Africa	243,300	-	243,300	295,000	-	295,000	51,700	21
Total Africa*	3,614,000	71,000	3,685,000	4,397,600	15,900	4,413,500	728,500	20
Americas	509,300	259,700	769,000	496,400	250,400	746,800	-22,200	-3
Asia and Pacific	3,615,200	280,100	3,895,300	3,551,900	278,300	3,830,200	-65,100	-2
Europe	3,057,000	18,200	3,075,200	4,362,600	28,800	4,391,400	1,316,200	43
Middle East and North Africa	2,898,500	65,400	2,963,900	2,675,400	64,100	2,739,500	-224,400	-8
Total	13,694,000	694,400	14,388,400	15,483,900	637,500	16,121,400	1,733,000	12

* Excluding North Africa.

Fig. 6 Refugee populations by UNHCR regions in 2015 [8]

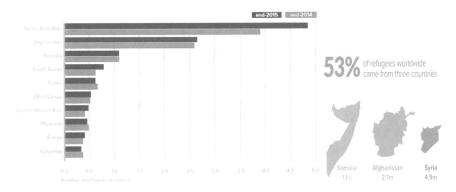

Fig. 7 Major source countries of refugees from 2014 to 2015 (end year) [8]

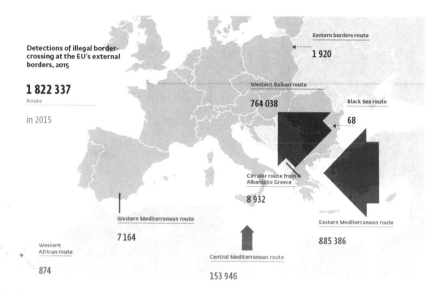

Fig. 8 Detected border crossings in 2015 [22]

The migrant flows from the Middle East and North Africa regions follow eight main passages to enter Europe [7]; routes and numbers are depicted in Figures 8, 9, and 10:

(a) The Eastern Mediterranean route from Turkey to Greece, with two branches: (a) the land route from Turkey to the region of Evros, (b) the sea route from Turkey to the Aegean islands of Samos, Lesvos, and Chios.
(b) The Central Mediterranean route from Libya to Italy and Malta.
(c) The Western Mediterranean route from Morocco to Spain, with three branches: (a) the route across the Gibraltar straits, (b) the route from the Spanish cities Ceuta and Melilla in Morocco to Spain, (c) the route via the Canary Islands.
(d) The Eastern European land route via the Russian Federation to Ukraine.
(e) The Arctic route via the Russian Federation to Finland and Norway.

Data from Frontex, Europol, and UNHCR shows that the main body of refugees arrive in Europe by sea; it is estimated that in 2015 over one million people arrived this way [8]. The Mediterranean Sea thus has become one of the most travelled seas of the twenty-first century. Unfortunately, it has also become one of the most dangerous. Thousands of people have drowned or gone missing during the sea travel on unsafe and overloaded vessels; the number of deaths and density of incidents for 2016 are shown in Figures 11 and 12. In fact, current data indicates that the risk of dying during the crossing is close to 2%. In that respect, it is also probably the most dangerous border on earth, considering that it divides countries not at war with each other [24].

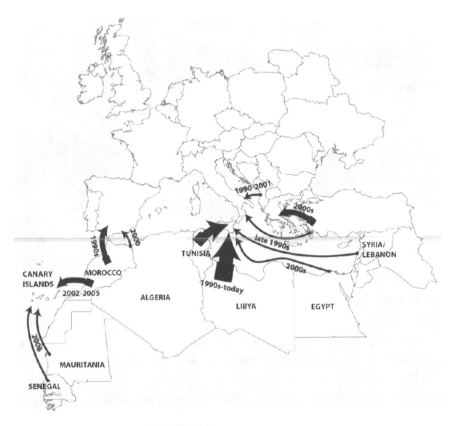

Fig. 9 Sea routes to Europe 1990–2015 [24]

In humanitarian crisis situations, clear demographic data is the key for decision-making. However, especially in cases of forced displacement, population groups are regularly in motion and their structure and human components are not stable. Furthermore, there is not one singular agency responsible for data collection; therefore, demographics are usually fractured and unreliable [5]. Some of the available composite data regarding both gender and age in Mediterranean arrivals is outlined in Figure 13.

Unlike other demographic data, projections regarding religious beliefs can only be based on circumstantial data. Since the UN and Eurostat do not include information on religion in their reports, the proportion of Muslims among refugees seeking asylum in Europe may only be calculated as a combination of their nationality (Figure 14) together with the religious composition of their home countries [26].

According to the CIA World Factbook 2016 [27], the religious makeup of the top nationalities in Mediterranean arrivals is predominantly Islamic, as depicted in Table 2.

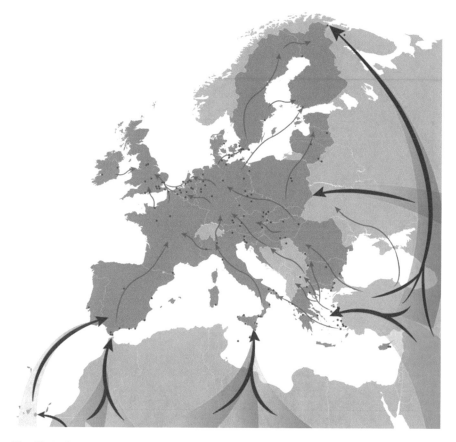

Fig. 10 Main routes to and within Europe [25]

Fig. 11 Dead and missing persons in 2018 [15]

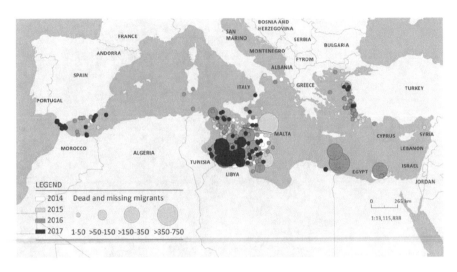

Fig. 12 Density of incidents [15]

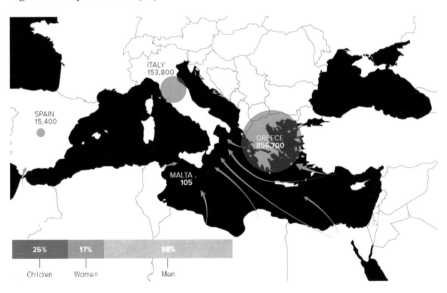

Fig. 13 Composite data regarding sex and age for Mediterranean arrivals in 2015 [8]

Consequently, it is reasonable to assume that 87% of all refugees arriving to Europe are Muslims. This estimation is significant, because religion is one of the most important drivers of animosity in hosting societies towards refugees [26].

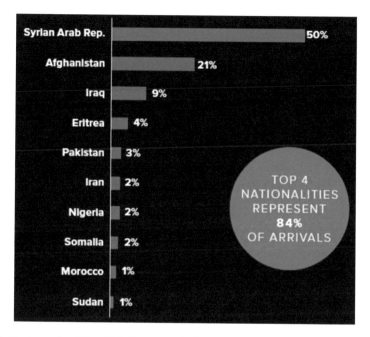

Fig. 14 Top ten nationalities of Mediterranean Arrivals in 2015 [8]

Table 2 Religious makeup in main refugee countries of origin [27]

Country of origin	Percentage of Muslim population
Syrian Arab Republic	87.00%
Afghanistan	99.70%
Iraq	99.00%
Eritrea	(Officially Muslim—no numerical data)
Pakistan	96.40%
Iran	99.40%
Nigeria	50.00%
Somalia	(Officially Muslim—no numerical data available)
Morocco	99.00%
Sudan	(Officially Muslim—no numerical data available)

Integration Issues

Forcibly displaced persons need security and acceptation, not exclusion and discrimination. However, it is very challenging to successfully integrate groups of people with already mixed national, religious, cultural, and social background into societies that are so heavily contrasting their own [8, 22].

Challenges of the Various Aspects of Integration

In order to avoid xenophobia, radicalization, and social fragmentation, policies should not be limited at addressing the most immediate and short-term humanitarian needs of refugees, such as registration, documentation, and temporary shelter. A long-term strategy must be developed, preferably at European level, to ensure homogeneity across member states and to prevent fragmentation within the European Union [28].

Such a strategy must include work opportunities, education, and social inclusion, defined as the progressive incorporation of displaced populations into their host country; it is a predominately qualitative concept and cannot be easily defined or measured. However, there are elements that can be identified as essential components of any successful assimilation process [5, 29]:

- *Legal*: legal residency status, equal access to justice, civil and political rights.
- *Economic*: equal economic opportunities, right to work, access to financial services.
- *Social*: right to social services (welfare, health care, and education), the absence of discrimination, participation in the social and cultural life, positive interaction with local communities.

Currently, there are significant deficits in one or more of these elements throughout Europe. Controversies in legal, social, and cultural issues exist in varying degrees in all hosting countries, depending on factors such as the level of relevant legislation, the mindset of local societies, or the pre-existence of similar ethnic groups from previous migrations; these controversies tend to become minor in countries hosting large numbers of refugees, while relying on a background of strong social infrastructure, such as Germany or Sweden [20, 30]; this does not apply to countries where the influx of new arrivals seriously threatens weak or nonexistent welfare systems. Health issues (e.g., interrupted vaccinations in areas of conflict, deteriorated refugee health, the spread of both communicable and non-communicable diseases) are being broadly ignored, especially in regard to children, despite the fact that they can easily affect the general population [22]; no comprehensive health plan has been developed thus far, and health institutions have remained for the most part silent [4]. Furthermore, prejudice and hostility towards Muslim population, originating back in the recession of the 1970s, has been evolving to an increased xenophobia and alarmism ever since. Research shows that, in the majority of European countries, anti-Muslim bias (Figure 15) is significantly more pronounced than anti-immigrant bias [31].

Social Impacts and Conflicts for the Hosting Societies

The large scale of this forced migration did evoke significant uneasiness and tension on a variety of topics within European countries [10]. The ideals of openness, free movement, and multiculturalism that formed the foundation of the EU had already

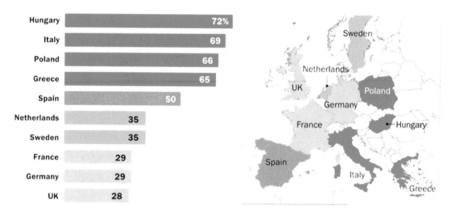

Fig. 15 Negative views of Muslims in European countries [6]

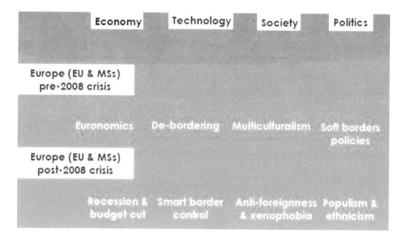

Fig. 16 The change of attitude and policies in pre- and post-2008 Europe [19]

suffered a major impact during the financial crisis of 2008. When the common economic policies failed to protect the Member States or assist in their recovery, the general attitude regressed back to restriction policies (Figure 16), thus consequently influencing the response to the refugee crisis [19].

The influx of refugees enhanced hostile feelings against "foreigners," which in turn radically increased the popularity of populist and nationalist extreme-right parties, abolishing at the same time the public support from government parties. These political reactions may have the protection of national identity and social coherence at their core, nonetheless they are enforcing them with acts of discrimination and often violence against the people regarded as intruders [10, 32]. On the other hand, the incidents of sexual harassment against local women during the 2016 New Year festivities in Germany and Sweden triggered feelings of insecurity, vulnerability, and anger. More importantly, the brutal terrorist attacks in

Fig. 17 Public opinion on the EU's handling of the refugee crisis [6]

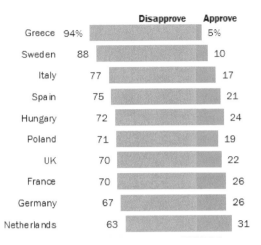

France and Belgium in 2015 and 2016 convinced both governments and public that radical Islamist groups have penetrated Europe using the mass movement of people as a cover-up, thus indiscriminately equating the refugees with terrorists and further stoking suspicion and hatred towards them [10].

These conflicts have led to increased border control, progressive isolation, and a strong trend against collective approach in several European countries and are causing disagreements to the various Member States. The escalating tension between EU members is aggravating already existing economic, social, and political schisms between countries and is threatening the general stability of Europe itself [21, 33]; this is reflected in the predominately negative public opinion pertaining to the EU's handling of the refugee crisis, illustrated in Figure 17 [6].

There is a very pragmatic risk that these problems will ultimately impede balanced refugee integration and will create migrants without roots, perpetually circulating from country to country and futilely seeking asylum [34]. There are currently at least 76,000–80,000 people stranded in Greece and the Balkans (Figure 18) without official status or the possibility to legally continue their travel [15].

International, European and National Legislation

Data shows that the current global migration trends will not abate easily [10]. However, the world in general and Europe in particular continue to remain unprepared for dealing with the waves of migration, both in humanitarian and in legislative terms; and while displaced people can demonstrate adaptability to circumstances, governments, organizations, and the public fail to do so [10]. Furthermore, quality of life—expressed as legal, economic, cultural, and social integration—needs to be clearly defined in a comprehensive, broadly accepted, and binding way [5, 35].

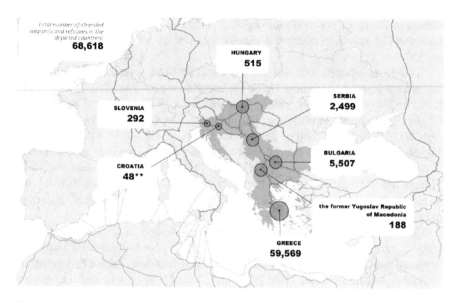

Fig. 18 Number of stranded people in Greece and the Balkans [15]

The Protection of Life in International Law

At the center of the refugee crisis lies the protection of all human life; it is a very complex issue that includes aspects such as protection of freedom and life from persecution due to race, religion, nationality, social or political preference, protection of children, sea rescue, and combat of human trafficking [19]. The existing relevant legislation is subdivided into four sections [36] namely: (1) International refugee law, (2) International human rights law, (3) International humanitarian law and the law of neutrality, and (4) International criminal law. These are not laws in the traditional sense, they are rather a sequence of statutes, declarations, convention proceedings, committee conclusions, and protocols; in that respect, they are better described as generally accepted guidelines, and therein also lies the difficulty of globally adopting and strictly enforcing them.

The European Legislation

In regard to the refugee issue, by the late 1990s Europe was moving towards the shared strategy of ending the "exilic bias" (i.e., of stemming the refugee flows through direct intervention in the conflict areas); furthermore, it had undertaken two more formal actions [3]: (a) the Dublin Convention, ruling the country of first entrance as the exclusively designated country for asylum application and (b) the

Schengen Group of countries, aiming at establishing shared border policies (e.g., movement of goods, services and people, visa policy, and one common external border).

In the years between 1995 and 2016, in the effort to manage, control, and restrain the refugee flows, the European Union has undertaken a number of further initiatives [19, 37, 38]. Among them, aiming to control the EU borders and regulate the refugee influx are the establishment of the agency for border control management (FRONTEX), the EU Border Surveillance System (EUROSUR), the Asylum, Migration and Integration Fund (AMIF), and the EU Naval Force Mediterranean (EUNAVFOR MED). Moreover, for dealing with the groups already within its borders the EU has enacted the European Asylum Support Office (EASO), the European Agenda on Migration, and first and second relocation plan; addressed to refugees that have entered in Greece, Italy, and Hungary [39].

These are more or less isolated and short-term attempts to achieve a measure of control over the crisis and not a serious effort to develop a multilateral strategy able to incorporate all aspects of the issue—legislative, economic, humanitarian, social, cultural or regarding security and integration. In order to achieve that, Europe has first to overcome an increasing distrust and apprehension towards a central European administration and the ensuing member states' strong tendency to act unilaterally in their own national interests [10, 19].

National Management Strategies: Implementation and Current Measures

Until 1995, every European country followed its own guidelines and legislation; these were in general restrictive and were focused on four actions [3]: (1) prevention of access (e.g., increased border security), (2) discouragement of asylum seekers (e.g., deficient accommodation conditions, limited rights and benefits), (3) efforts to accelerate refugee determination procedures, and (4) repatriation of failed applicants.

Unfortunately, disagreements between member states regarding a plethora of issues have further deteriorated the situation to the point that countries presently implement conflicting measures to address the refugee crisis. Actions to restrict access range from increased controls to barbed-wire fences on external borders (e.g., Hungary, Slovenia, and Bulgaria), to temporary border controls even within the Schengen area. Some countries reject the predetermined quota of refugees, impose restrictions on access to asylum processes, and family reunification (e.g., Austria, Denmark, Finland, Germany, and Sweden) or refuse to receive non-Christian groups in their territories. Whereas some countries (e.g., Norway, Sweden) issue temporary work permits for highly skilled refugees together with the possibility of education for minors, other countries refuse them these opportunities, while simultaneously unduly prolonging the asylum processes [8, 9, 28].

To conclude, actions taken so far both at the EU level and at the national level are short-term, fragmented, antagonistic, incomplete, unclear, and ineffective. In order to avoid unpredictable consequences in the future, essential is the development and implementation of a long-term strategy focusing on social and economic integration for the refugees, by providing equal opportunities in work, education, health care, and housing, without at the same time compromising the coherence as well as the safety and security of the hosting societies [3, 28].

Critical Assessment of the Current Situation and the Available Solutions

The need for permanent and stable resolution regarding all forcibly displaced persons—refugees, IDP's, and stateless persons—is clear and irrefutable for everyone involved. Beyond full legal recognition, there are three established permanent solution alternatives, pursued mostly by the UNHCR [5, 8, 10]. More specifically: (1) voluntary repatriation: the return to their homes is obviously the most popular among refugees, (2) resettlement: relocation to a third country is an option whenever refugee needs cannot be met in the country of first asylum, and (3) local integration: this entails legal, economic, cultural, and social inclusion, and it encompasses a permanent home, the ability to sustain a livelihood, the economical contribution to the host country, and the lack of intolerance or unfairness.

So far, the achievement of a reliable solution remains an unresolved challenge. The incessant warfare in the whole MENA region in general suggests that repatriation cannot be considered as an option yet; only about 200,000 out of the 7.5 million displaced people decided to repatriate voluntarily in 2015 [8]. Furthermore, no willing relocation countries exist, since most of them have already exhausted their available resources. Thus, the only feasible and reasonable solution for the foreseeable future is local integration [38].

It is established that a responsible and successful assimilation should form part of the middle ground between the undisciplined and uncontrolled entry into a host country with its inherent risk to safety and the total debarment due to an irrational and unjustifiable xenophobia. And since integration commences with first reception, it is essential that hosting countries must develop extended reception capacities, improve the efficiency of national asylum systems, and increase refugee recognition rates [38, 40]. However, even in Europe conditions are not always ideal, since the load is carried predominantly by those countries with the weaker economies, while a significantly large number of refugees are still enclosed in emergency facilities suffering progressively adverse conditions [10, 41].

Long-Term Policies and Framework

The Problem of Shelter

Considering the fact that refugees must by definition be addressed as a vulnerable social group, a critical and central aspect of all first reception, temporary accommodation, and full inclusion is shelter in the broader sense of sanctuary from adversities and dangers. The concept of "adequate standard of living" is already included in the EU Reception Conditions Directive (2013/33/EU, article 18) in connection to the actual waiting period of the asylum application process; this concept should be expanded to the full duration of a person's stay within a country's jurisdiction.

Adequate and decent housing must be ensured in order to protect the life and health of people, prevent sexual or gender-based violence, and address the specific needs of children [9]. The first step towards this objective is the establishment of emergency admission, registration, and assistance centers [40]. Beyond that essential initial stage, the important need for long-term accommodation creates the

Table 3 Advantages and disadvantages of enclosed settlements and open housing

Advantages	Disadvantages
1. *Enclosed settlements*	
• Enhanced safety and protection • Better identification and estimation of population numbers • Improved monitoring of health status • Easier organization of basic services (e.g., distribution of goods, vaccinations) • Centralized support systems • Easier planning of future options (e.g., repatriation) • Improved economy of scale in the provision of services • Better communication possibilities • Better skills coaching (e.g., language teaching)	• Increased risk of disease outbreaks due to overcrowding • Strong dependence on external support • Diminished autonomy • Social isolation • Possible degradation of the surrounding environment • Possibility of security problems within the camp
2. *Open housing*	
• Higher persons mobility • Better access to external jobs • Use of existing infrastructure (e.g., public health system) • Faster reconstruction of economic substance	• Increased effort to access en masse the total refugee population • Difficulty in monitoring health needs or emergencies • Diminished access to relief programs (e.g., food aid) • Risk of destabilization in the local community and frictions between local residents and refugees • Possible decline of whole urban districts into ghettos

ongoing controversy between the two currently predominant alternatives: planned camps vs. housing in private apartments.

Each of the two alternatives favors both significant advantages and serious disadvantages (Table 3), thus making the choice between the two a difficult one [36, 42]. The suitability and choice between the two alternatives cannot be absolutely determined, as it depends heavily on individual circumstances: number of refugees in relation to the absorbing capacity of local communities, demographic constitution of each refugee group, ethnic, cultural, and religious compatibility and tolerance level between refugees and local residents, specific political situation, and equilibrium within the receiving nations.

However, one general conclusion that can be drawn with relative certainty is that planned camps should be predominantly short-term oriented, whereas individual housing within the boundaries of local societies is more suitable as a long-term solution on the path to full social integration [36, 42].

Refugee Settlements: Existing Guidelines and Design Criteria

Natural and man-made disasters are not a rare occurrence and as a result, the various agencies involved in the management of such extreme situations (e.g., UNHCR, US Army, The Red Cross, Doctors without Borders, etc.) have all developed sets of practical standards to optimize the relief operations. Most significant among them is the "Humanitarian Charter and Minimum Standards in Disaster Response"; it is the result of the Sphere Project, initiated in 1997 as a teamwork between international non-governmental organizations and continuing to evolve ever since [14].

General Design and Construction Principles

An inadequately designed refugee settlement can become an ailing environment, both literally and figuratively, therefore optimal planning and organization are essential, in order to minimize corrective actions, make management easier and more cost-effective, and achieve the most efficient allocation of land and resources [36, 42]:

- Basic needs that have to be addressed include shelter, essential healthcare, nutrition, water and sanitation. Other issues involve resource logistics, camp coordination and camp management (CCCM), non-food items (NFI, e.g., clothing, bedding, etc.), telecommunications and security [43].
- Initial design should focus on optimal camp size and density to avoid overcrowding—both for health and security reasons—flexibility to adapt to changing requirements and advance planning for all seasons and weather conditions—for health and environmental reasons and to achieve maximum cost-effectiveness [42, 44].
- Health is always one of the major concerns in any area where large groups of people congregate for longer time periods. Some of the most common health

hazards include pollution of surface and groundwater, contamination of the environment, development of breeding areas for disease carrying vectors, and presence and spread of insects and rodents. They are attributed to poor sanitation and waste water management, insufficient or inadequate garbage reception points, dust in the air or smoke, and they need to be addressed during the design phase [36].

- In emergency situations, changes in the traditional demographic structure of groups are to be expected; they include the absence of men as traditional family care-takers and transition of this role to female family members with subsequent increased risks for their safety, significant numbers of unaccompanied children, and increased numbers of older, sick, injured, or disabled people. These changes create additional special circumstances and gender considerations and need to be taken into account [36].

- The first step prior to the planning of a refugee settlement should be an environmental baseline study to (a) determine the status of the current environmental situation, (b) detect any possible sensitive issues, such as environmentally protected sites that the camp should distance itself from, (c) calculate the use of local resources, (d) appraise the actual impact of the settlement to the environment, all with the scope to mitigate or minimize as much as possible the temporary and permanent adverse effects [44].

- Integral part of the initial planning should be a comprehensive exit strategy, not necessarily in regard to the duration of stay for the consecutive cycles of refugee groups, but more importantly in terms of an end-of-life approach, i.e., the fate of the facilities after they have concluded their avail. In that respect, a simple decommissioning cannot be considered the most efficient solution; a more creative approach should be incorporated within the original concept [36].

Site Selection and Planning

Optimal site selection is the foundation of any successful settlement. Even though it might seem difficult, or at times outright impossible, every effort should be made so that the chosen plot will satisfy the majority of characteristics listed in Table 4 [42, 44]. As a general rule, however, overestimation of potential needs or adversities is preferred to underestimation [36].

Site Organization, Infrastructure, and Services

In designing the spatial organization of the settlement, the factors to be taken into account include minimum space allocation per person, individual space requirements for each installation, minimum distances required between various uses, and easy accessibility of all services (Table 5).

Equally important are qualitative aspects such as security factors, social structure, cultural traditions, relationships, and vulnerable groups within the population [42].

Table 4 General prerequisites for the eligibility of settlement building sites

1. *Topography and size*
• Plot: Almost flat, above flood levels, without extreme surface variations
• Ideal slope: 2–4% and < 10%
• Soil: Absorb surface water easily
• Rocky subsoil: To be avoided
• Groundwater table: >300 m below the surface
• Plot size: Satisfy all the space allocation criteria and additional free area reserves
• Plot outline: Allow for a low-density design
• Area required for social and communal functions: Not to be underestimated
2. *Water resources*
• Location: Close to a source of good, potable water
• Drill-well construction: Only after a detailed hydrological survey, and only when no other option is available
• Transportation of water by tank trucks: To be avoided when possible
• Quantities of water: Sufficient to cover the settlement's demands; allow for the probable excess use of water
3. *Public utilities*
• Location within or very close to the public utilities grid (i.e., water, sewage, and electricity)
4. *Accessibility*
• Adequate road infrastructure
• Proximity to towns, markets, hospitals, and other national public services
• Access to public transportation
5. *Security*
• Location away from international borders (over a 50 km radius), conflict areas, or other sensitive areas
• Areas with extreme climatic conditions to be avoided
6. *Environment and vegetation*
• Site: Dispose sufficient vegetation to provide shade, wind protection, decreased soil erosion, and dust generation, improved micro-climate
• Trees: Not to be an impediment to construction
• Soil: Allow for small-scale gardening and production of vegetables
7. *Land rights*
• Inhabitants of the settlement: Exclusive rights to use the plot
• Public land made freely available: Good choice
• Legal and traditional land rights or uses: Not to be violated

The objective of these emergency settlements should extend beyond creating a simple housing space to protect from the elements and safeguard life and health; ideally, they should recreate and enhance a sense of privacy and security for the displaced people [42, 44]. In regard to the administrative and community services available within the settlement (Table 6), the possible requirements can vary depending on a number of factors; among others, number of people housed in the settlement, particular needs or specific nature of the population, planned duration of stay.

Moreover, according to the total size and population of the settlement, some functions—usually administrative—should be centrally located and easily acces-

Table 5 Infrastructure requirements for emergency settlements

---
1. *Electrical supply and distribution*

- #1: Security lighting, access lighting and equipment operation
- #2: Individual living quarters; minimum requirements are light and one power outlet
- #3: Electrical power to be provided for heating and cooling of the individual units
---
2. *Water supply and distribution*

- Potable water: For drinking and cooking
- Potable water: Preferably for all other settlement operations; if quantities are not adequate, non-potable water can be used for cleaning and bathing
- Connection to the local public water system: Preferred option
- A gravity-fed distribution system within the settlement recommended
- Absolute minimum capacity of the system: 20 L/day/person for whole population
- Further water requirements: Possible firefighting; grey water or rainwater use is also recommended
- Water treatment facility: Recommended for maximum resource efficiency
- Rainwater collection system: Is also desirable
---
3. *Fire prevention*

- Adequate firebreaks (i.e., distance between structures)
- Location of potential fire hazards (e.g., fuel storage areas) into consideration
- Fireplugs: Located in appropriate spots everywhere within the camp
---
4. *Access roads and parking*

- Access roads within the settlement: Designed to address every possible daily activity or emergency situation
- Road surfaces: All-weather, appropriate for heavy-duty vehicles
- Specific design requirements (e.g., minimum dimensions, turning radius) to be applied
---

Table 6 Typical administrative and community services requirements for emergency settlements

1. Administration	• Administrative office • Registration office • Social services office	• Archive room • Reception and waiting area
2. Health	• Medical center • Pharmacy	• Infirmary
3. Food	• Kitchen—Food preparation area • Dining area	• Food storage (cold, frozen and dry goods)
4. Sanitation	• Latrines • Washing and bathing areas • Laundry area	• Laundry supply storage • Garbage disposal areas
5. Community	• Teaching area • Community area—Congregation area	• Religious areas—Prayer rooms
6. Warehousing	• Non-food items storage.	• Distribution center

sible for all, others—usually those regarding personal hygiene—are best to be decentralized to accommodate smaller groups of people [36, 44].

Quantitative Standards for Spatial Allocation and Services

Specificquantitative standards regarding space allocation, services, and infrastructure are detailed in Table 7 [36]. All the standards quoted above are evidently not legally binding; rather, they are the result and combination of both theoretical analysis and practical experience gained on the field. These guidelines are designed to establish a baseline level of protection, comfort, and dignity for those forcibly displaced persons, while at the same time maximizing time and resource efficiency and minimizing possible adverse effects. For those reasons, it is strongly advisable that a serious effort be made for their application, if not improvement, as a whole.

The Case Study of the Accommodation Center Project

At this point it must be once more stressed that EU and its immediate neighbors has done and is doing nothing to comply with the aforementioned international standards. The radical example of the Tripoli Zoo in Libya, used as a temporary detainee center since 2011, showcases the often inhumane housing conditions imposed on people that have lost everything and in the majority of cases through no fault of their own. These grossly violating basic human rights conditions are not solely occurring in the so-called third world countries, but in member states of the EU as well. There, accommodation facilities range from a minority of large, conforming to guidelines, and professionally managed centers all the way to the majority of small, improvised, inadequately equipped, and badly controlled "hotspots," usually housed in abandoned buildings or warehouses and often lacking even the most basic goods like shelter or sanitation [45]. For the refugees, and in view of their anticipated extent of stay in the area, these actions should—as mentioned before—include the provision of decent housing, medical services, and opportunities in education or training and work. On the other hand, hosting countries should be assisted in relieving the strain on their own basic infrastructure regarding health and education, as well as on the connectivity of their social fabric [46, 47].

The Accommodation Center Concept

Everything that has been elaborated on insofar is in its majority a theoretical approach to a very complex and multifaceted issue. It is in light of the above extensive theoretical review and in search for feasible and applicable alternatives well within the realm of realization that this accommodation center concept has been developed. It concerns the creation of a prefabricated multi-functional model settlement to act as an intermediary hospitality center for the refugee population in any host country. This project obviously does not have the ambition to address and

Table 7 Quantitative standards for emergency settlements

Space allocation	
Land	• 30—45 m²/person
Sheltered space	• 350 m²/person/min. Ceiling height: 2 m
Fire break space	• 50 m wide area between shelters for every 300 m built area
Roads and walkways	• 20–25% of entire site
Open space	• 15–20% of entire site
Site gradient	• 1–5% (ideally: 2–4%)
Water	
Water supply	• Min. 20 L/person/day
Water tap stand	• 1 per 80 persons
Water distance	• Max. 200 m from household unit (optional: 100 m) • No further than a few minutes' walk
Water pipes	• Depth 40–60 cm to avoid damage from surface activities • Areas with low temperatures: Depth 60–90 cm to avoid frost
Sanitation	
Latrines	• 1 per 20 persons/optional: 1 per family, separate latrines
Latrine distance	• Max. 50 m—Min. 6 m from household unit • Close enough to facilitate use, but far enough to prevent smells and pest issues
Shower	• 1 per 50 persons, separate shower areas for men and women
Refuse container	• 1 × 100 L/50 persons
Communal refuse pit	• Size: 2 m × 5 m × 2 m, 1 per 500 persons.
Health	
Health center	• 1 per 20,000 persons, optional: 1 per settlement
Referral hospital	• 1 per 200,000 persons
Food	
Nutritional value	• 2100 kcal/person/day
Food quantity	• 36 tonnes/10,000 people/week
Feeding center	• 1 per 20,000 persons • Optional: 1 per settlement
Warehousing	
Storage area	• 15–20 m²/100 persons, optimal: individual refugee storage
Commodity distribution area	• 1 per 5000 persons

(continued)

Table 7 (continued)

Space allocation	
Communal services	
School	• 1 per 5000 persons
Market place	• 1 per 20,000 persons, optional: 1 per settlement
Administration	
Administration offices	• As appropriate • Includes all administrative functions
Security	
Lighting	• As appropriate • Emphasis on priority areas (latrines, public areas) and security
Security post	• As appropriate
Security fencing	• As appropriate • Depending on individual circumstances and security issues

resolve the whole problem of social integration in its entirety; merely, it is intended as what is perceived a necessary intermediary step between first reception and full integration or eventual relocation and even repatriation [48].

All relevant data points to the fact that homogenization of populations with significant pre-existing barriers regarding the language, religion, culture, and ethics, if it is not well-prepared, cannot be achieved without considerable turbulences for both sides. And if for the local residents this groundwork is mostly limited in a general understanding of the existing differences, the neutralization of illogical fears, and the development of tolerance and acceptance, for the refugee "newcomers" it entails issues much more practical and urgent in nature, considering that they directly influence their ability to survive in dignity in their new surroundings. The transitional accommodation center aims to create a secure, unthreatening environment where the feeling of safety can be restored, the integration obstacles can be in part or in whole removed or at least smoothed out and the acclimatization to the new circumstances can be achieved in relatively controlled conditions.

The main characteristic of the proposed settlement is its compact design, considering that it demands a comparatively small land plot of only 4000—5000 m^2 to accommodate 500 people; the reasonable size of the plot and the moderate number of residents help to create a more user-friendly, village-like atmosphere—without however compromising the existing guidelines regarding space allocation or excluding any of the desirable functions [36]. These functions of the settlement have been specifically selected to form part of the general strategy of facilitating the prospective inclusion in the hosting society as well as providing the stepping stones for a decent future standard of living. In addition to the provision of safe, equipped with all essential amenities and reasonably comfortable living quarters, which is the most basic function of the accommodation center, the supplementary five services that have been deemed essential for the success of the project include:

Administration

Administrative services are organized to provide assistance, handle complications, and perform all relevant duties in regard to four distinct sets of issues; more specifically, the objective is (a) to coordinate and regulate the short- and long-term management of the center, (b) to execute the recording, filing, and archiving of all relevant refugee data, (c) to disentangle and resolve legislative and bureaucratic issues regarding asylum procedures and immigration laws, residency permits, or even repatriation, and (d) to resolve problems and facilitate in every way the interaction between the refugee population and the local society. These tasks may seem overly ambitious; however, given the relatively small number of people residing within the center at any given time, they can be accomplished competently and with a significant probability of success.

Health

The accommodation center is equipped with a small medical center with the purpose to (a) perform initial medical screenings and general health checks, (b) provide the necessary medications, vaccinations, and consistent treatment protocols of possible pre-existing diseases, (c) attend to problems such as undernourishment or exhaustion, and (d) deal with small, everyday medical emergencies. The scope is to address moderate health issues in a consistent and organized way, without the risk of interrupted or inappropriate treatments while at the same time without unduly burdening the regular public health infrastructure if not absolutely imperative.

Food

To ensure the correct nourishment of the refugees is deemed a most important aspect of their stay within the center, especially in regard to the more vulnerable groups among them and until their good health is fully restored. Individual kitchen facilities increase construction cost while at the same time presenting an a serious risk for accidents; further than that, many of the residents may not be in a position to prepare food to themselves (e.g., unaccompanied children, elderly or disabled people, etc.). Therefore, the provision of food in (a) adequate quantities, (b) decent quality, and (c) nutritional value as prescribed by the guidelines is an essential amenity of the accommodation center.

Education

The educational services pivot around two main axes. One is the learning of the local language, since this will remove one of the most significant barriers of inclusion;

at the same time, the refugees can get acquainted with the ethics and customs of their new home country. The other axis is training in basic working skills for those who lack any, or most importantly, assistance in the official recognition of existing skills and knowledge according to the host country laws; this might include university degree validation processes or licensing examinations. Part of this educational mechanism is also the sharing of knowledge among people, given that local instructors will cooperate with accordingly qualified members of the refugee population in order to facilitate the learning process and additionally remove possible traces of distrust.

Work Opportunities

The final step before integration in the hosting society is preparation for job placement, considering that decent work according to individual qualifications is the essential means for unaided and dignified sustenance and, given time, evolution and prosperity within the society. The theoretical components of this process involve support in recognizing competencies, workshops on interviewing skills, and assistance in creating an effective CV. The practical components include part-time jobs, at first within the settlement and then in the immediate area, in cooperation with the local authorities and residents.

In addition to all of the above, the center's operation principles include the provision that the inhabitants will assist in the management and day to day operations (e.g., cleaning, teaching and sharing knowledge, assisting in food distribution, partaking in maintenance work, etc.), each one according to his or her specific qualifications, talents, or capabilities, but with no exceptions whatsoever besides impairing health problems. This allocation of work among the inhabitants, already recommended by existing guidelines, will assist in creating and preserving a sense of ownership and responsibility for the refugees, but also achieve a reduction in operation cost and human resources required [36].

Design and Construction

As it is outlined in the general layout plan of the settlement (Figure 19, courtesy of Icon Architecture), the main body of the building is located at the front part of the plot, facing the access road and entrance to the settlement; it houses all the centralized functions. More specifically: (1) administration offices, (2) medical ward, (3) reception and waiting area, serving both administrative and medical needs, (4) food preparation area, (5) food storage area, (6) multi-functional rooms, mostly for educational or religious use, and (7) indoors central gathering area, designed as a common area for sitting, dining, communicating, and socializing.

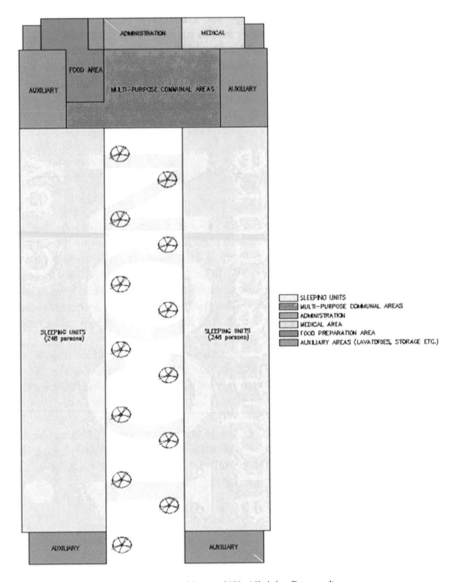

Fig. 19 Accommodation center general layout [49], All rights Reserved)

It must be noted that the areas designated for the use of the refugees are deliberately characterized by increased versatility, in order to accommodate a wider variety of functions, both as dictated by the general everyday needs of the inhabitants but also according to the specific demands arising from whatever demographic structure they might have in every separate occasion.

Towards the back of the plot and away from the road and the main entrance unfold the two wings of housing quarters; this spatial organization ensures an increased level of privacy combined with a lower level of general disturbances (e.g., from traffic, noise, etc.). These quarters incorporate: (1) individual living modules, (2) latrines and washing areas, and (3) laundry areas and cleaning supplies storage rooms.

The living modules are the place where the refugees can retreat in, not only to sleep, but also to rest or enjoy some private moments; they cannot be very generous in dimensions due to the general size restrictions, they are however within the minimum space allocation guidelines and they do provide enough room for the basic equipment (i.e., beds, closets, tables, and chairs). Between the two wings and freely accessed from the indoors gathering area lies an open-air inner courtyard, protected from the elements with light fabric tents and dedicated to communal outdoor activities. Sleeping modules facing the inner courtyard are reserved for the most vulnerable and in need of protection residents, such as unaccompanied children, single women, youths, elderly or disabled people, and households with predominantly female or underage members. Accordingly, sleeping modules facing the outside are assigned to the more capable and self-reliant members of the community.

In regard to utilities, the electrical lines, water pipes, and sewer pipes serving the complex run along a walkway created by the back sides of the two rows of opposite facing sleeping modules in each wing. These lines are open and exposed for easy installation, control, and repairs; the walkway however is fully enclosed and accessible only to the maintenance crew. An internal service road surrounds the complex and provides easy approach to every part of it, both for everyday needs (e.g., garbage collection) and for any other exigent circumstances (e.g., medical emergencies, fires, etc.).

The building is assembled from simple shipping containers. The external walls and roofs are constructed from insulated aluminum panels, 6–10-mm thick, according to the use and to the climatic conditions of every individual location. The whole settlement is constructed in the factory as prefabricated units and then transported and assembled on-site according to the plans. It is important that the building site should be selected according to the existing guidelines, since this ascertains strong reductions in construction time and cost. Due to the light construction, the site work needed includes minor earthworks for ground leveling, a light foundation for the containers, the construction of the essential infrastructure, i.e., the main lines for electricity, water, and sewer, as well as the internal roadwork. The described mode of construction offers a significant number of advantages such as: (1) very short on-site construction time, (2) uncomplicated installation, (3) optimal maintenance, (4) above average insulation, (5) seismic safety, and (6) optimal relationship between quality and cost. Moreover, the project design is also incorporating environmentally friendly technologies to promote sustainability, especially in the energy and waste management sector.

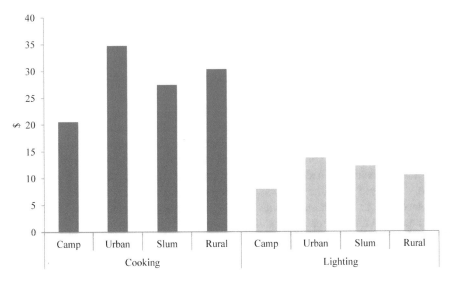

Fig. 20 Per capita annual spending on energy by forcibly displaced people in different settings [50]

Renewable Energy Sources

One of the significant issues related to the daily operation and services of an emergency settlement, and one of the most easily overlooked, is energy consumption. As a result, an investment on energy infrastructure is usually not a central concern in emergency humanitarian relief. However, studies show that in only one year, 2014, energy use from forcibly displaced people globally amounted to the equivalent of almost four million tonnes of oil being burned, while the cost of energy for cooking and lighting per year per family of five was calculated at $200 minimum (Figure 20); this amount adds up to an unwarranted global total cost of over two billion dollars in the same year [50]. Moreover, the consumption of this amount of fossil fuel equivalents results in significant CO_2 emissions, with all negative consequences to the environment.

In order to address these issues, at least in part, the accommodation center electrical installation planning includes the provision for the positioning of a photovoltaic panel system on the roof of the building. Solar energy is maybe the least demanding of the renewable energy sources in terms of installation, service, and operation, combined with a relatively low construction cost. Furthermore, if the construction budget is adequate and the location is suitable, there is the option of additionally installing one or more small wind turbines as a further contribution to the energy demands of the settlement. In view of the fact that this particular accommodation center project is designed to be by default situated on the outskirts of urban areas, the produced electrical energy can be fed directly on the grid and used according to the operational needs. However, the technological progress in respect to energy storage devices (i.e., batteries) will, in the very near future, make the construction of similar hosting facilities feasible even in off-grid areas by ensuring their energy autonomy.

Waste Management

Even though it tends to be disregarded as well, waste management is a challenging issue, especially in cases where considerable numbers of people live and circulate in a relatively dense space for longer periods of time; in such cases the volume and weight of garbage can easily reach significant amounts [36]. The general categories of waste produced in any such case are (a) grey water, i.e., water from bathing and laundry, (b) toilet waste, and (c) solid waste, e.g., organic remains from food, packaging materials, papers, etc. The present accommodation center is designed to interconnect with its neighboring communities, consequently waste disposal can effect through their public infrastructure; and yet, waste management should include more than simple garbage collection and removal.

Concerning grey water, treatment systems are available on the market, their installation however entails considerable cost and complicity, as it necessitates separate drainage lines and as a result they cannot be regarded as an option for the purposes of the specific project. Together with toilet waste, grey water is to be disposed in the public sewer network; in the rare cases where no such network exists, septic tanks will be constructed at the periphery of the site and at its lowest level to collect the waste and to be emptied at regular intervals. Nevertheless, as a partial compensation and even though rainwater cannot be characterized as waste per se, an independent system of rainwater collection, comprising of roof gutters and rainwater tanks, will be installed. This system will both resolve the issue of surface drainage that can potentially create problems regarding soil erosion or possible flooding and also make the center more resource efficient by utilizing the collected water for auxiliary purposes (e.g., plant irrigation, cleaning of solar panels or outdoor areas, firefighting), thus reducing the consumption of potable water significantly.

Nonetheless, the main by-product of human activity, both in volume and in treatment issues, is solid waste. As regards the sub-category of organic waste, the usual method of management is transport to the local landfills; yet, as a preferred alternative, it can be utilized for composting within the settlement; this is an optimal way for the disposal of food remains, with the additional advantage of potentially becoming an income generating initiative. Even though this process requires some technical knowledge, there are small systems easily available that can be operated without difficulties and with little skill; to facilitate the process, and besides the appropriate garbage bins dispersed throughout the settlement, a separate collection area for kitchen waste will be constructed in close proximity to the food preparation area. Included therein will also be vats specifically assigned for the collection of cooking oils and fat.

In relation to all other forms of refuse, and based on the concept of waste separation to facilitate and promote the objective of reuse or recycling, the proposed method is installation of several green, i.e., recycling spots [51]. Specially appointed and enclosed areas, similar to the plan in Figure 21, and located on the periphery of the site will accept the appropriate individual trash receptacles, each designated for a different material category and demarcated accordingly with a different color (Figure 22), including, as a minimum, the following: (1) paper and cardboard (e.g., packaging materials), (2) plastic, (3) glass, (4) metal, and (5) electrical and electronic equipment. Furthermore, optionally: (1) mixed packaging (e.g., tetra pack), (2) hazardous household waste (e.g., cleaning products packaging, solvents, etc.), (3) wood and timber, (4) batteries, (5) garden waste, and (6) fabrics and clothing items.

Special provisions are made for medical waste (e.g., used syringes, needles, bandages, expired medicines, etc.), that can be potentially hazardous. For that purpose, a securely enclosed collection area equipped with biohazard containers will be constructed as an attachment to the medical center; that waste will be disposed separately by the medical personnel only. To conclude, instigated by the awareness that waste generation deriving from the center's operation is the sector with the most impact to the surrounding environment, the aim is to manage the substantial amount of waste in the most efficient and environmentally friendly way [52].

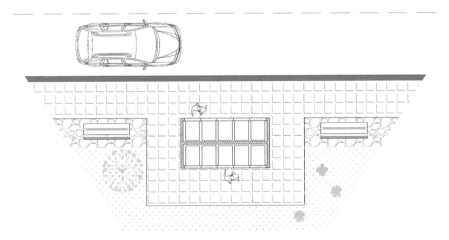

Fig. 21 Indicative layout of a neigborhood "island" green spot with 800 m^2 area size [51]

Fig. 22 Examples of color coded recycling bins [48, 51]

Techno–Economic Study

Duration of construction and cost of construction are integral parts of any project. Consequently, an initial assessment for the determination of these two factors is also included in the accommodation center project; meantime, an effort has been made to ascertain that this preliminary approach is valid, even if subject to unforeseeable factors.

Concerning the duration of construction, there are two distinct phases: (a) construction of the units in the factory and (b) assembly time on-site. Factory construction time is estimated between 14 and 18 weeks, depending on timely orders, size of the factory, and punctuality of payments. Within this timeframe, the site work and utilities infrastructure can also be completed. Finally, on-site assembly time is estimated between 12 and 14 weeks, mostly depending on weather conditions and work crews' coordination. Moreover, a reasonable amount of time for the transportation of the prefabricated units must be allowed in addition to these timeframes.

With respect to the total construction cost, as outlined in Table 8, it is tentatively calculated at 1.7–2 million dollars, under the provision that the accommodation center will be interconnected with the local infrastructures. However, it must be stressed that this is only a very rough estimate of cost. A number of factors—most important among them the country of installation—are affecting all areas, building, site work, utilities, and peripherals, they are responsible for critical outlay variations and as a consequence they will influence the total construction expenditure significantly. Some of these variable costs and their causes and impacts are included in Table 9.

On the other hand, when evaluating the construction cost, significant factors must be taken into consideration: (a) the initial cost breakdown per capacity is estimated only for one group of inhabitants; however, the center will accommodate consecutive group of refugees and therefore this cost will decrease significantly throughout the center's operation cycle (e.g., assuming a conservative two groups/year the cost breakdown for the first year drops to $1700–$2000/inhabitant), and (b) the initial construction cost should be evaluated in relation to the total projected utilization and life cycle of the project as is elaborated further.

Table 8 Estimated construction cost range

1. *Building*	
• Reception area	• Individual living modules for ~500 people
• Administration offices	• Latrines and washing areas
• Medical ward	• Laundry rooms
• Food preparation and storage area	• Storage rooms
• Multi-functional rooms	
• Central gathering area	
2. *Site work*	
• Site clearing and fill	• Roadwork
• Foundations	• Landscaping
3. *Utilities*	
• Water installation	• Photovoltaic installation
• Sewer installation	• Green spot installation
• Electrical installation	• Rainwater collection installation
4. *Peripherals*	
• Building permit	• Furniture and equipment
• Transport	• Miscellaneous
• Assembly	
Estimated total cost range (VAT not included)	$ 1,700,000—2,000,000
Cost breakdown: • As per built space: $600—$700/m^2 • As per capacity: $3.400—$4.000/inhabitant	

Differentiated Second Phase Utilization and End-of-Life Approach

In every project with the ambition to be sustainable, a crucial and integral, even if easily ignored aspect that must be taken into consideration from the initial steps in planning onwards is the end-of-life approach, i.e., the attempt to establish the optimal way to conclude its circle of existence. In the present case, after it has fulfilled its original purpose, and instead of being demolished or fall into disuse, the accommodation center can conveniently be modified and further utilized in different capacities; in that respect, the prefabricated metal construction is an important feature that contributes to the versatility of the project. By adding or removing modules the building can be easily enlarged or reduced in size, according to the future needs; likewise, partition walls can be removed or erected to modify the footprint of rooms and utility lines, being open and accessible, can be easily moved or extended. Even the elevations can undergo a transformation, by being painted, stuccoed, or even clad with different materials to support and enhance the new function. Taking the layout of the center into consideration and with the condition that only minor alterations should be required, the potential alternative uses can include: (1) educational facilities (e.g., school, training center), (2) shopping center,

Table 9 Construction cost fluctuation factors

Variable costs	Causes and impacts
1. Construction site characteristics	• Soil quality (e.g., sandy or rocky soil) • Gradient • Distance from existing infrastructures

2. Materials cost	• Global fluctuations of raw materials cost
	• Local material prices
	• Impact on:
	  – Factory construction cost
	  – Utilities installation cost

3. Labor cost	• Local cost of labor:
	  – In country of factory
	  – In country of assembly
	• Impact on:
	  – Factory construction cost
	  – Assembly cost
	  – Utilities installation cost

4. Transportation cost	• Distance from factory to building site • Accessibility of building site
5. Additional features	• Septic sewer • Wind turbines • High security fencing, etc.
6. Contractor fees	• Building contractor fees range: 10–20% – The project however might be assigned to local, national, or army technical services
7. Taxes, etc.	• Local taxes related to the construction • VAT percent variations • Currency exchange rates

(3) office complex for civil authorities or private professionals and practitioners (e.g., lawyers, doctors, engineers, etc.), (4) primary health center, (5) homeless shelter and welfare services, (6) assisted living center for elderly or disabled people, and (7) accommodation for students (i.e., dormitories).

Even in the case that, for some reason the occupation of the original lot is no longer feasible or desired (e.g., the location is not fit for the prospective use, the landowners want the property returned to them, etc.), the settlement can be further utilized still. Entailing a cost much lower than traditional demolition, and again due to its construction, the building can be dismantled and transported to a different location to be reassembled and used anew; moderate reparations will of course be required, but yet again they are not comparable to the construction cost and complexity of a new facility of this size. The settlement can retain its original role as a refugee accommodation center, or assume any other alternative use as mentioned above.

At this point it must be mentioned that, since the concept of the project could be applied to any major catastrophe, there is the additional option of creating an inventory of such settlements, warehoused and ready to be assembled at any time and in any place needed. Even if it seems far-fetched, this suggestion is feasible and should be taken into consideration from any country where the statistical probability of crisis situations entailing destruction of properties and requiring emergency shelter (e.g., hurricanes, floods, wildfires, earthquakes, etc.) is higher than regular.

When it has ultimately reached its end of life, the building can then be permanently dismantled and the majority of construction materials (e.g., steel, aluminum, and drywall) can be easily recycled, thus contributing to both environmental and economic efficiency. Even the cost of building site reinstatement is small; given that little initial work regarding foundations was demanded, no major earthwork is required to restore the site to its original condition. To conclude, and irrelevant of the chosen end-of-life alternative, the refugee accommodation center remains a project both efficient and compensatory all the way through from its commencement to its termination.

Conclusions

Mass movements of populations, either planned and deliberate or urgent and impulsive, are not new to societies. They have happened before and they will continue happening for as long as the underlying causes, such as economic insecurity, food scarcity, conflicts, or natural disasters—those induced by climate change included—do not cease to exist. Among them, the refugees, fleeing their homes out of fear for life and safety, are the ones mostly in need of support and assistance; however, they are also grossly neglected in terms of official recognition and legal status. Induced from those mass movements, the ever-increasing demand for humanitarian assistance in emergency situations initially led to the formation the Red Cross in 1863 and was further addressed with the foundation of the United Nations High Commission for Refugees; even today, UNHCR remains the predominant actor and retains the highest authority on refugee issues [53]. As everyday goods become scarce in increasingly large parts of the world and both natural and man-made disasters occur with disturbing frequency, the distinct scientific branch of humanitarian logistics has evolved in order to satisfy the need for efficient management of goods, materials, information and human resources, also combined with advance preparation, data processing, and use of expert knowledge, thus ensuring the success of any humanitarian relief operation. Unfortunately, problems affecting societies globally do not seem to be abating and humanitarian relief, however successful, will continue to be in high demand. The focus of this paper was to examine the current refugee crisis affecting the European and MENA region. The scope was to develop a compact yet also versatile refugee accommodation center, based on the principles of sustainability and suitable to address part of the needs arising from the current or any other similar disaster situation [54].

The analysis of historical data from the nineteenth century onwards has illuminated the fact that this seemingly unprecedented tide of forcibly displaced people is not a novelty in the European continent. In fact, within the last 200 years, Europe has experienced no less than six different, bigger or smaller waves of forced migration within its boundaries, each with its own distinct characteristics and ensuing problems. The final, current refugee wave was instigated in part by the incessant poverty in North-Central Africa and South-Central Asia, but more importantly, by the violent armed conflicts in Eritrea, Iraq, Afghanistan, and most recently Syria. The deliberate annihilation of infrastructures in the areas of conflict, the dire conditions in the countries of first asylum in the Middle East, as well as the fall of the Gaddafi regime in Libya and the ensuing opening of sea routes in the Central Mediterranean all contributed to the rapid inflation of the refugee movement towards the European region.

However, even though Europe has always been the theater of internal displacements, only in recent years has it become such a highly desirable hosting region, thus unwillingly assuming the role of a strong actor, compelled to provide assistance rather than receiving it from others, but also reluctant to act; therein lies the significant distinguishing factor between the current and all former forced migration situations, and not in the size of the refugee populations. Despite the negative climate within Europe and taking into consideration the continuing instability in the MENA region—and the so far dormant displaced populations (e.g., Palestinians) potentially influenced into seeking asylum in Europe, the prognosis for the future is anything but optimistic; and with conflict generally becoming the pattern in world affairs, forced displacement will not seize to exist globally.

Even though it is not always clear or complete, the analysis of available statistical data has helped to highlight distinct features and particularities of the current refugee wave [55]. Concerning religion, one of the main causes of anxiety in Europe, the available data is only circumstantial, since no agency includes relevant information in their reports. Still, the estimated percentage is around 87%. To conclude, it is indisputable that a long-term presence of large refugee groups with all their various characteristics and dissimilarities is inevitable in Europe. This commands the development of a long-term strategy with an aim towards full social integration. To achieve this predominately qualitative goal, this strategy must incorporate and address a variety of issues of legal, economic, and social nature. Nonetheless, at present there are significant deficiencies and discrepancies in all these areas among hosting countries, depending on the pre-existing state of their relevant infrastructure, their political and economical standing and the mindset of their citizens.

The expected tension ensuing by the large scale of forced migration has been amplified beyond measure on the fertile ground of post-2008 Europe. After the financial crisis of 2008, European nations regressed to a state of introversion. Distrust for a European central governing authority, increased border control, xenophobia and ethnicism flourished, overturning the ideal of an open Europe. These feelings, further supported by the recent brutal terrorist attacks, found a voice to express themselves in increased hostility and discrimination against refugees, thus ignoring all positive aspects their arrival might entail [29]; additionally, they resulted in a strong trend against collective actions among countries. Measures taken so far are disjointed or sporadic, prioritizing border control, maritime surveillance, search and rescue operations, and the establishment of bureaucratic rather than humanitarian supporting structures. As a result, no reasonable and complete, but also fair and humanitarian crisis management strategy yet exists [56]. Nevertheless, and regardless of divergent political backgrounds and refugee policies, the demand for a durable and solid solution is undeniable. As a result, hosting countries, whether they realize it or not, are facing the dilemma of either showing indifference and thus allowing a forced penetration, probably in the outskirts of society with all the entailing risks, or developing strategies for controlled and managed social and economical integration.

One of the major issues and also integral part of a successful social inclusion process is the problem of shelter. Adequate housing is essential for the protection of life and health, elimination of sexual and gender-based violence and regard for the specific needs of children, not only during initial emergency admission but also all the way into full assimilation. In respect to that, the current discussion between the two existing alternatives, planned camps or open housing, has been analyzed and the advantages and the disadvantages of both options presented. It has become clear that the choice is not influenced only by the specific characteristics of each proposal, but it is also further dependent on individual circumstances, e.g., the absorbing capacity of communities, the demographic profile of refugee groups, the social tolerance and compatibility levels and, not least of all, the stage of integration. Furthermore, the suitability of any emergency response settlement strongly relies on a set of criteria regarding design and construction, site selection and planning, site organization, infrastructure and services, as well as spatial needs. Unfortunately, this does not apply to the majority of the reception centers established in countries throughout Europe, where living conditions are most often unacceptable and, in some cases, violate even the not basic human rights. Within this context, and as part of a focused uniform plan of action, the paper has elaborated on the development of a compact accommodation and hospitality center made of shipping containers, to function as an intermediary stage in adaptation between initial reception of the forcibly displaced and full social integration. This project aims at maximizing the respect for human rights and values and also at minimizing the impact on society and on the environment. The main objective is to create a secure, unthreatening environment capable to restore the sense of safety and dignity of the forcibly displaced people, while at the same time removing the barriers of language, religion, culture, and ethics and, in parallel, relieving the strains on local infrastructures, mostly on health and education. This goal has led to the integration of administrative, medical, educational, job related, religious, and social functions within the settlement, together with the full inclusion and participation of the inhabitants in the operations and proceedings of the accommodation center. Regarding the environmental sustainability, the main issues addressed are optimal land utilization, renewable energy use, and waste management programs. Creating added value for the "raw" material (shipping containers) and prolonging the unit's life span by enabling transformation and change of use, transportation and reuse, and finally end-of-life dismantlement and recycling also lie within this scope. Furthermore, this project is versatile enough to be expanded and adapted for the implementation on further social groups in need of support.

To conclude, it can be maintained that, in order to facilitate the whole assimilation process between forcibly displaced and local populations, and taking advantage of the existing policy void, this concept is attempting to create a novel intermediary integration phase not previously considered. Moreover, the results could serve as a useful tool for governments and organizations to better plan ahead and respond fast and efficiently not only in regard to the actual refugee crisis, but also in any possible humanitarian disaster situation, including those emanating as consequences of climate change.

References

1. UNHCR: Figures at a Glance. [online] http://www.unhcr.org/.html (2016). Accessed 16 Nov 2016
2. Cozzolino, A.: Humanitarian logistics – cross-sector cooperation in disaster relief management. [online] http://www.springer.com/gp/book/9783642301858 (2012). Accessed 3 Nov 2016
3. Robinson, V.: The changing nature and European perceptions of Europe's refugee problem. Geoforum. [online] **26**(4), 411–427 https://doi.org/10.1016/0016-7185(95)00042-9 (1995). Accessed 5 Nov 2016
4. The Lancet: Adapting to mitigation as a planetary force (editorial). The Lancet [online] **386**(9998), 1013 http://thelancet.com/journals/lancet/article/PIIS0140-6736(15)00190-7/fulltext (2015). Accessed 2 Nov 2016
5. UNHCR: UNHCR Statistical Yearbook 2014, 14th edn. [online] http://www.unhcr.org/ (2015). Accessed 11 Nov 2016
6. Pew Research Center: 5 Facts about the Muslim population in Europe. http://www.pewresearch.org/fact-tank/2016/07/19/5-facts-about-the-muslim-population-in-europe/ (2016). Accessed 19 Nov 2016
7. Parkes, R.: Asylum flows to the EU: blip or norm? [online] http://www.iss.europa.eu/de/publikationen/detail/article/asylum-flows-to-the-eu-blip-or-norm/ (2016). Accessed 31 Oct 2016
8. UNHCR: UNHCR Global Trends 2015. [online] http://www.unhcr.org/ (2016). Accessed 11 Nov 2016
9. European Union Agency for Fundamental Rights: Asylum and migration into the European Union in 2015. [online] http://fra.europa.eu/en/publication/2016/asylum-and-migration-european-union-2015 (2016). Accessed 11 Nov 2016
10. Vimont, P.: Migration in Europe: bridging the solidarity gap [online]. http://carnegieeurope.eu/2016/09/12/migration-in-europe-brid-ging-the-solidarity-gap-pub-64546 (2016). Accessed 5 Nov 2016
11. Suarez-Orozco, M.: Immigration and Migration: Cultural Concerns. In: Smelser, N., Baltes, P. (eds.) International Encyclopedia of the Social & Behavioral Sciences. 1st edn., p. 7211. [online] http://www.sciencedirect.com/science/reference-works/9780080430768 (2001). Accessed 6 Nov 2016
12. Amnesty International: People on the move. [online] https://www.am-nesty.org/en/what-we-do/people-on-the-move/ (2016). Accessed 2 Nov 2016
13. Bascom, J.: Refugees: geographical aspects. In: Smelser N., Baltes, P. (eds.) International Encyclopedia of the Social & Behavioral Sciences. 1st edn., 12895–12901. [online] http://www.sciencedirect.com/science/referenceworks/9780080430768 (2001). Accessed 6 Nov 2016

14. Black, R.: Refugees and displacement. In: Kitchin, R., Thrift, N. (eds.) International Ency-clopedia of human geography. 1st edn. vol. 9, pp. 125–129 [online] http://www.sciencedirect.com/science/referenceworks/9780080449104 (2009). Accessed 6 Nov 2016
15. International Organization for Migration: International Organization for Migration. [online] http://www.iom.int/ (2016). Accessed 8 Nov 2016
16. Thomas, A., Kopczak, L.: From logistics to supply chain management: the path forward in the humanitarian sector. [online] http://www.fritzinstitute.org/PDFs/WhitePaper/FromLogisticsto.pdf (2005). Accessed 3 Nov 2016
17. Ergun, Ö., Keskinocak, P., Swann, J., Heier Stamm, J., Villareal, M.: Relief operations: how to improve humanitarian logistics. [online] http://analytics-magazine.org/relief-operations-how-to-improve-humanitarian-logistics/ (2010). Accessed 3 Nov 2016
18. Forced Migration Review: Delivering the goods: rethinking humanitarian logistics. [online] http://www.fmreview.org/logistics.html (2003). Accessed 6 Nov 2016
19. Attina, F.: Europe faces the Immigration Crisis. Perceptions and Scenarios. [online] http://www.transcrisis.eu/wp-content/uploads/2015/11/new-repaper10.pdf (2015). Accessed 5 Nov 2016
20. Achilli, L.: Tariq al-Euroba: displacement trends of Syrian asylum seekers to the EU. [online] http://cadmus.eui.eu/handle/1814/38969 (2016). Accessed 7 Nov 2016
21. Heisbourg, F.: The strategic implications of the Syrian refugee crisis. [online] https://www.iiss.org/en/publications/survival/sections/2015-1e95/survival%2D%2Dglobal-politics-and-strategy-december-2015-january-2016-522a/57-6-02-heisbourg-12d7 (2015). Accessed 6 Nov 2016
22. Frontex: Risk analysis for 2016. http://frontex.europa.eu/as-sets/Publications/Risk_Analysis/Annula_Risk_Analysis_2016.pdf (2016). Accessed 11 Nov 2016
23. Angeli, D., Triandafyllidou, A.: Europe. In: Migrant smuggling data and research: A global review of the emerging evidence base - International Organization for Migration. [online] https://publications.iom.int/books/migrant-smuggling-data-and-research-global-review-emerging-evidence-base (2016). Accessed 7 Nov 2016
24. Fargues, P., Bonfanti, S.: When the best option is a leaky boat: why migrants risk their lives crossing the Mediterranean and what Europe is doing about it. [online] http://cadmus.eui.eu/handle/1814/33271 (2014). Accessed 6 Nov 2016
25. Europol: Migrant smuggling in the EU. [online] https://www.europol.europa.eu/publications-documents/migrant-smuggling-in-eu (2016). Accessed 8 Nov 2016
26. Pew Research Center: The future of the global Muslim population - Region: Europe. [online] http://www.pewforum.org/2011/01/27/future-of-the-global-muslim-population-regional-europe/ (2011). Accessed 19 Nov 2016
27. USA Central Intelligence Agency: The World Factbook. [online] https://www.cia.gov (2016). Accessed 19 Nov 2016
28. Hvenmark-Nilsson, C.: The European refugee crisis: the need for long-term policies and lessons from the Nordic Region. [online] Center for Strategic and International Studies. https://www.csis.org/analysis/european-refugee-crisis-need-long-term-policies-and-lessons-nordic-region (2015). Accessed 2 Nov 2016
29. International Monetary Fund: The refugee surge in Europe: economic challenges. Staff discussion notes. [online] http://www.imf.org/external/ns/cs.aspx?id=353 (2016). Accessed 2 Nov 2016
30. The Expert Council of German Foundations on Integration and Migration: Immigration Countries: Germany in an International Comparison. 2015 Annual Report. [online] http://www.svr-migration.de/en/press/press-expert-council/svr-releases-2015-annual-report/ (2016). Accessed 19 Nov 2016
31. Strabac, Z., Listhaug, O.: Anti-Muslim prejudice in Europe: a multilevel analysis of survey data from 30 countries. Soc. Sci. Res. [online] **37**(1), 268–286 http://www.sciencedirect.com/science/article/pii/S0049089X07000142 (2008). Accessed 3 Nov 2016

32. European Commission: Refugee crisis in Europe. Humanitarian Aid and Civil Protection. [online] http://ec.europa.eu/echo/refugee-crisis_en (2016). Accessed 19 Nov 2016
33. Klinke, I. The geopolitics of European (dis)union. Polit. Geogr. [online] **37**, 1–4. http://www.sciencedirect.com/science/article/pii/S0962629813000681 (2013). Accessed 6 Nov 2016
34. Kuusisto-Arponen, A., Gilmartin, M.: The politics of migration. Polit. Geogr. [online] **48**, 143–145. http://www.sciencedirect.com/science/article/pii/S0962629815000566 (2015). Accessed 6 Nov 2016
35. Behm, A.: Why the refugee crisis is a strategic issue. [online] http://www.aspistrategist.org.au/why-the-refugee-crisis-is-a-strategic-issue/ (2015). Accessed 6 Nov 2016
36. UNHCR: Emergency Handbook. [online] https://emergency.unhcr.org/ (2016). Accessed 8 Nov 2016
37. Tsourdi, E., De Bruycker, P.: EU asylum policy : in search of solidarity and access to protection. [online] http://cadmus.eui.eu/handle/1814/35742 (2015). Accessed 6 Nov 2016
38. UNHCR: UNHCR Global Appeal 2016–2017 – Europe regional summary. [online] http://www.unhcr.org/ (2015). Accessed 11 Nov 2016
39. European Union Agency for Fundamental Rights: Handbook on European law relating to asylum, borders and immigration. [online] http://fra.europa.eu/en/publication/2013/handbook-european-law-relating-asylum-borders-and-immigration (2014). Accessed 11 Nov 2016
40. Clayton, J.: UNHCR chief issues key guidelines for dealing with Europe's refugee crisis. [online] http://www.unhcr.org/en-us/news/latest/2015/9/55e979 3b6/unhcr-chief-issues-key-guidelines-dealing-europes-refugee-crisis.html (2015). Accessed 6 Nov 2016
41. European Commission: Humanitarian aid report. Special Eurobarometer 434. [online] http://ec.europa.eu/public_opinion/archives/eb_special_439_420_en.htm (2015). Accessed 11 Nov 2016
42. Medecins Sans Frontiers: Public Health Engineering in Precarious Situations. 2nd edn. [online] http://refbooks.msf.org/msf_docs/en/public_health/public_health_en.pdf (2010). Accessed 7 Nov 2016
43. Haddow, G., Bullock, J., Coppola, D.: International disaster management. In: Introduction to Emergency Management, 5th edn., pp. 263–304 [online] http://www.sciencedirect.com/science/book/9780124077843 (2014). Accessed 6 Nov 2016
44. USA Department of Air Force: Refugee Camp Planning and Construction Handbook. Air Force Handbook 10–222, vol. 22. [online] http://www.dtic.mil/docs/citations/ADA423967 (2000). Accessed 8 Nov 2016
45. Vaughan-Williams, N.: "We are not animals!" Humanitarian border security and zoopolitical spaces in EUrope. Polit. Geogr. [online] **45**, 1–10. http://www.sciencedirect.com/science/article/pii/S0962629814000900 (2015). Accessed 6 Nov 2016
46. Pierini, M., Hackenbroich, J.: A bolder EU strategy for Syrian refugees. [online] Carnegie Europe. http://carnegieeurope.eu/2015/07/15/bolder-eu-stra-tegy-for-syrian-refugees/ided (2015). Accessed 6 Nov 2016
47. Sarkis, J.: Models for compassionate operations. Int. J. Prod. Econ. [online]. **139**(2), 359–365. http://www.sciencedirect.com/science/article/pii/S0925527312002630 (2012). Accessed 5 Nov 2016
48. Papadaki, S.: Refugees: The Humanitarian Logistics of a crisis situation. Master Dissertation Thesis, MSc in Environmental Management and Sustainability, School of Economics, Business Administration & Legal Studies, International Hellenic University, February 2017 (2017)
49. Icon Architecture: © "Xenios Zeus" refugee accommodation project. Planning and design. All rights reserved (2015)
50. Lehne, J., Blythe, W., Lahn, G., Bazilian, M., Graftham, O.: Energy services for refugees and displaced people. Energy Strat. Rev. [online]. **13–14**, 134–146 http://www.sciencedirect.com/science/article/pii/S2211467X16300396 (2016). Accessed 18 Nov 2016
51. Operational Programme Environment and Sustainable Development. Guide for the planning, organization and operation of green spots. [online] http://www.epper.gr/el/Pages/Default.aspx (2015). Accessed 19 Nov 2016

52. European Union Joint Research Centre: Life cycle indicators for resources, prod-ucts and waste: Waste management. Institute for Environment and Sustainability. [online] https://ec.europa.eu/jrc/en/publication/eur-scientific-and-technical-re-search-reports/life-cycle-indicators-resources-products-and-waste-waste-management (2012). Accessed 14 Nov 2016

53. Mason, E.: Resolving refugee problems: an introduction to the Executive Committee of the United Nations High Commissioner's Programme and its documentation. J. Gov. Inf. [online] **27**(1), 1–11 http://www.sciencedirect.com/science/article/pii/S1352023799001562 (2000). Accessed 18 Nov 2016

54. United Nations: In Safety and Dignity: Addressing Large Movements of Refugees and Migrants. Secretary-General's Report. A/70/59 [online] http://refugeesmigrants.un.org/secretary-generals-report (2016). Accessed 19 Nov 2016

55. European Asylum Support Office: Annual report on the situation of asylum in the EU 2015. [online] https://www.easo.europa.eu/information-analysis/annual-report (2016). Accessed 14 Nov 2016

56. European Parliament Directorate-General for External Policies: Current challenges for interna-tional refugee law, with a focus on EU policies and EU co-operation with the UNHCR. [online] http://bookshop.europa.eu (2013). Accessed 11 Nov 2016